Mathematics for Technical Trades

Nigel P. Cook

Upper Saddle River, New Jersey
Columbus, Ohio

Editor in Chief: Stephen Helba
Senior Acquisitions Editor: Gary Bauer
Editorial Assistant: Natasha Holden
Production Editor: Louise N. Sette
Production Coordinator: Holly Henjum, Carlisle Publishers Services
Design Coordinator: Diane Ernsberger
Cover Designer: Bryan Huber
Production Manager: Brian Fox
Marketing Manager: Jimmy Stephens

This book was set in Times Roman by Preparé. It was printed and bound by Banta Book Group. The cover was printed by Phoenix Color Corp.

Copyright © 2004 by Pearson Education, Inc., Upper Saddle River, New Jersey 07458.
Pearson Prentice Hall. All rights reserved. Printed in the United States of America. This publication is protected by Copyright and permission should be obtained from the publisher prior to any prohibited reproduction, storage in a retrieval system, or transmission in any form or by any means, electronic, mechanical, photocopying, recording, or likewise. For information regarding permission(s), write to: Rights and Permissions Department.

Pearson Prentice Hall™ is a trademark of Pearson Education, Inc.
Pearson® is a registered trademark of Pearson plc
Prentice Hall® is a registered trademark of Pearson Education, Inc.

Pearson Education Ltd.
Pearson Education Singapore Pte. Ltd.
Pearson Education Canada, Ltd.
Pearson Education—Japan

Pearson Education Australia Pty. Limited
Pearson Education North Asia Ltd.
Pearson Educación de Mexico, S. A. de C.V.
Pearson Education Malaysia Pte. Ltd.

10 9 8 7 6 5 4 3 2 1
ISBN: 0-13-045269-6

To Dawn, Candy, and Jon

Books by Nigel P. Cook

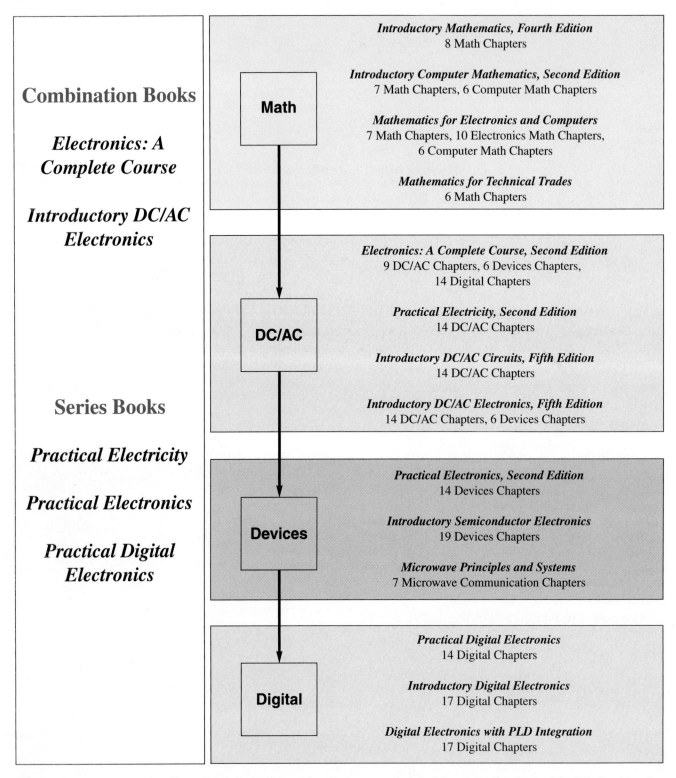

Combination Books

Electronics: A Complete Course

Introductory DC/AC Electronics

Series Books

Practical Electricity

Practical Electronics

Practical Digital Electronics

Math

Introductory Mathematics, Fourth Edition
8 Math Chapters

Introductory Computer Mathematics, Second Edition
7 Math Chapters, 6 Computer Math Chapters

Mathematics for Electronics and Computers
7 Math Chapters, 10 Electronics Math Chapters,
6 Computer Math Chapters

Mathematics for Technical Trades
6 Math Chapters

DC/AC

Electronics: A Complete Course, Second Edition
9 DC/AC Chapters, 6 Devices Chapters,
14 Digital Chapters

Practical Electricity, Second Edition
14 DC/AC Chapters

Introductory DC/AC Circuits, Fifth Edition
14 DC/AC Chapters

Introductory DC/AC Electronics, Fifth Edition
14 DC/AC Chapters, 6 Devices Chapters

Devices

Practical Electronics, Second Edition
14 Devices Chapters

Introductory Semiconductor Electronics
19 Devices Chapters

Microwave Principles and Systems
7 Microwave Communication Chapters

Digital

Practical Digital Electronics
14 Digital Chapters

Introductory Digital Electronics
17 Digital Chapters

Digital Electronics with PLD Integration
17 Digital Chapters

For more information and a desk copy of any of these textbooks by Nigel P. Cook, call 1-800-228-7854, visit the Prentice Hall website at www.prenhall.com, or ask your local Prentice Hall representative.

Preface

INTRODUCTION

Mathematics is interwoven into the very core of science, and therefore an understanding of mathematics is imperative for anyone pursuing a career in technology. As with any topic to be learned, the method of presentation can make a big difference between clear comprehension and complete confusion. Employing an "integrated math applications" approach, this text reinforces all math topics with extensive applications to show the student the value of math as a tool; therefore, if the need is instantly demonstrated, the tool is retained.

OUTLINE

After 14 textbooks, 21 editions, and 19 years of front-line education experience, best-selling author Nigel Cook has written *Mathematics for Technical Trades* as a complete math course for technology students.

Chapter 1   Whole Numbers and Fractions
Chapter 2   Decimal Calculation
Chapter 3   Positive and Negative Numbers
Chapter 4   Exponents and the Metric System
Chapter 5   Algebra, Equations, and Formulas
Chapter 6   Geometry and Trigonometry

SUPPLEMENTS

- Instructor's Manual (ISBN 0-13-049227-2)
- PH TestGen (ISBN 0-13-114157-0)
- Companion Website: http://www.prenhall.com/cook

An Extensive Bank of Integrated End-of-Section and End-of-Chapter Technical Trade Questions

38. Landscaping Determine angle x for the lawnmower shown.

39. Construction The *plummet* was used by the Egyptians to tell whether stones were level. Describe how this could be used.

40. Automotive What triangles are formed within the automobile jack shown?

41. Electronics The size of a television screen is measured diagonally as shown. From the measurements given, determine the length of side x.

42. Carpentry From the measurements given, determine the unknown length (x) of the roof frame.

43. Automotive The drum shown is used to store waste oil. What is its volume?

44. Plumbing How many cubic feet of volume does the water heater shown have?

45. Automotive An engine's *compression ratio* is the ratio of a cylinder's bottom-dead-center (BDC) volume to its top-dead-center (TDC) volume. What is the compression ratio of the engine cylinder shown?

Web Site Questions

Go to the Web site http://www.prenhall.com/cook, select the textbook *Mathematics for the Technical Trades*, select this chapter, and then follow the instructions when answering the multiple-choice practice problems.

The change in input (Δx) value will always be positive, but the change in output value can be either positive or negative based on whether y increased or decreased. To explain this point, take a look at the following two examples.

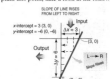

When a slope rises from left to right, the slope is said to be positive, and so y should be a positive value.

When a slope falls from left to right, the slope is said to be negative, and so y should be a negative value.

Any linear equation containing two variables can be written in the **slope-intercept form (SIF)**.

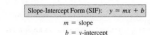

$m =$ slope
$b =$ y-intercept

Slope-Intercept Form
A form used to express the slope of a line in which $m =$ slope and b indicates the y-intercept,
SIF: $y = mx + b$.

In this appropriately named form, m represents the slope ($\Delta y/\Delta x$) and b the y-intercept. The main characteristic of an equation in slope-intercept form is that it is solved for y, which means that y should be by itself on the left side of the equation.

■ **EXAMPLE**

Referring to the graph, determine the equation for the line and give it in slope-intercept form.

■ *Solution:*

Slope rises from left to right and so y is positive.

$$\text{Slope} = \frac{\Delta y}{\Delta x} = \frac{2}{1} = 2$$

y-intercept $= +2$

The equation of the line in slope-intercept form is:

(SIF) $y = 2x + 2$

■ **EXAMPLE**

Convert the following equation to standard form (STF) and slope-intercept form (SIF), and then graph the line.

$$3y - 2x = 9$$

Conversational Writing Style, Visually Descriptive Layout, and Humanistic Vignettes

■ **Solution:**

Borrow 1 from 6 and convert it to $\frac{3}{3}$.

$$6\frac{1}{3} - 2\frac{2}{3} = 4\frac{1-2}{3}$$
$$6\frac{1}{3} = 5 + \left(\frac{3}{3}\right) + \frac{1}{3} = 5\frac{4}{3}$$
$$5\frac{4}{3} - 2\frac{2}{3} = 3\frac{4-2}{3} = 3\frac{2}{3}$$

Steps:
Subtract whole numbers; $6 - 2 = 4$.
Because fractions have the same denominator, subtract numerators. You cannot subtract 2 from 1. Borrow whole number from 6, and convert it to a fraction $\left(1 = \frac{3}{3}\right)$. Add 3 pieces to 1 piece to get 4 pieces, or 4 thirds.
Perform subtraction.

■ **TECHNICAL TRADE APPLICATION: BANKING**

The interest rate on a car loan has been reduced $2\frac{1}{4}$ percent from $6\frac{3}{16}$ percent. What is the new interest rate?

■ **Solution:**

$$6\frac{3}{16} - 2\frac{1}{4} = 4\frac{3-4}{16} \leftarrow \text{You cannot subtract 4 from 3.}$$
$$5 + \frac{16}{16} + \frac{3}{16} - 2\frac{1}{4} =$$
$$5\frac{19}{16} - 2\frac{1}{4} = 3\frac{19-4}{16} = 3\frac{15}{16}$$

SELF-TEST EVALUATION POINT FOR SECTION 1–4

Now that you have completed this section, you should be able to:

■ **Objective 10.** Show how to subtract proper and mixed fractions.

Use the following questions to test your understanding of Section 1–4.

1. $\frac{4}{32} - \frac{2}{32} = ?$
2. $\frac{4}{5} - \frac{3}{10} = ?$
3. $\frac{6}{12} - \frac{4}{16} = ?$
4. $15\frac{1}{3} - 2\frac{4}{12} = ?$
5. $5\frac{15}{36} - 2\frac{31}{36} = ?$

Use the following technical trade questions to test your understanding of practical applications of Section 1–4.

1. **Carpentry** What is the difference in height between the two cabinets shown?

2. **Drafting** Determine the missing dimension (x).

3. **Surveying** What is the difference in height between the two drains shown?

4. **Electrical** What is the power rating difference between a $\frac{1}{4}$ watt resistor and a $2\frac{1}{2}$ watt resistor?

5. **Machining** The diameter of a steel cylinder is reduced $\frac{14}{1000}$ inch from its original $\frac{750}{1000}$ inch. What is the cylinder's new diameter after being machined?

6. **Plumbing** The inside diameter of a steel pipe measures $\frac{5}{8}$ inch (i.d. $= \frac{5''}{8}$) while the outside diameter measures $\frac{12}{16}$ inch (o.d. $= \frac{12''}{16}$). What is the wall thickness of the pipe?

SECTION 1–4 / SUBTRACTING FRACTIONS

Integrated Math Applications, Worked-Out Example Section, and Self-Test Evaluation Points

Calculator Sequence Examples, Extensive End-of-Chapter Problems, and Margin Term Definitions

CALCULATOR KEYS

...nction

...calculators have a factor function that
...n factored with respect to all its variables.

factor (expression)

...ct the common factor from $(64x - 8y)$.
...actor $(64x - 8y)$
...)
...factor $(16a - 4)$
...)

To reverse the factoring process and convert $9(3x - 1x)$ back to the original equation $27x - 9x$, simply reverse the arithmetic steps. Because we began by dividing both numbers by the common factor, simply do the opposite and multiply both values within the parentheses by the factor 9, as follows:

$$9(3x - 1x)$$
$$= 27x - 9x$$

Steps:
$9 \times 3x = 27x,$
$9 \times 1x = 9x$

To make any equation with parentheses easier to solve, always begin by removing the parentheses before doing any other operation.

This process of removing parentheses, or multiplying through, is called *distributing*, and **the distributive law of multiplication** is as follows.

$$\boxed{a \times (b + c) = (a \times b) + (a \times c)}$$

Distributive Law of Multiplication: The same result is obtained regardless of whether you multiply a whole or multiply each part of the whole.

CALCULATOR KEYS

Name: Expand function

Function: Some calculators have a reverse factoring function that removes the parentheses.

Expand (expression)

Examples:

Press keys: Expand $(8(8x - y))$
Answer: $64x - 8y$
Press keys: Expand $(4(7a + 2b - 4c))$
Answer: $28a + 8b - 16c$

SECTION 5–2 / TRANSPOSITION

ILLUSTRATED TOUR OF TEXTBOOK FEATURES

Contents

1
Whole Numbers and Fractions 2
Vignette: There's No Sleeping When He's Around 2
Outline and Objectives 3
Introduction 4
1-1 Whole Numbers and Counting 4
1-2 Representing Fractions 6
1-3 Adding Fractions 9
1-4 Subtracting Fractions 17
1-5 Multiplying Fractions 20
1-6 Dividing Fractions 23
1-7 Canceling Fractions 25
Review Questions 27

2
Decimal Calculation 30
Vignette: An Apple a Day 30
Outline and Objectives 31
Introduction 32
2-1 The Decimal Point 32
2-2 Calculating in Decimal 41
Review Questions 59

3
Positive and Negative Numbers 62
Vignette: Blaise of Genius 62
Outline and Objectives 63
Introduction 64
3-1 Expressing Positive and Negative Numbers 65
3-2 Adding Positive and Negative Numbers 67
3-3 Subtracting Positive and Negative Numbers 70
3-4 Multiplying Positive and Negative Numbers 73
3-5 Dividing Positive and Negative Numbers 75
3-6 Order of Operations 78
Review Questions 81

4
Exponents and the Metric System 84
Vignette: Not a Morning Person 84
Outline and Objectives 85
Introduction 86
4-1 Raising a Base Number to a Higher Power 86
4-2 Powers of 10 94
4-3 Weights and Measures 100
Review Questions 116

5
Algebra, Equations, and Formulas 118
Vignette: Back to the Future 118
Outline and Objectives 119
Introduction 120
5-1 The Basics of Algebra 120
5-2 Transposition 125
5-3 Substitution 144
5-4 Equations and Graphing 152
5-5 Rules of Algebra—A Summary 163
Review Questions 165

6
Geometry and Trigonometry 168
Vignette: Finding the Question to the Answer! 168
Outline and Objectives 169
Introduction 170
6-1 Basic Geometric Terms 170
6-2 Plane Figures 173
6-3 Trigonometry 185
6-4 Solid Figures 195
Review Questions 200

Appendices

A Answers to Self-Test Evaluation Points 205
B Answers to Odd-Numbered Problems 215

Index

Mathematics for Technical Trades

Whole Numbers and Fractions

There's No Sleeping When He's Around!

Carl Friedrich Gauss was born April 30, 1777, to poor, uneducated parents in Brunswick, Germany. He was a child of precocious abilities, particularly in mental computation. In elementary school he soon impressed his teachers, who said that mathematical ability came easier to Gauss than speech.

In secondary school he rapidly distinguished himself in ancient languages and mathematics. At 14, Gauss was presented to the court of the duke of Brunswick, where he displayed his computing skill. Until his death in 1806, the duke generously supported Gauss and his family, encouraging the boy with textbooks and a laboratory.

In the early years of the nineteenth century, Gauss's interest was in astronomy, and his accumulated work on celestial mechanics was published in 1809. In 1828, at a conference in Berlin, Gauss met physicist Wilhelm Weber, who would eventually become famous for his work on electricity. They worked together for many years and became close friends, investigating electromagnetism and the use of a magnetic needle for the measurement of current. In 1833 they constructed an electric telegraph system that could communicate across Göttingen from Gauss's observatory to Weber's physics laboratory. (This telegraph system of communication was later developed independently by U.S. inventor Samuel Morse.)

Gauss conceived almost all of his fundamental mathematical discoveries between the ages of 14 and 17. There are many stories of his genius in his early years, one of which involved a sarcastic teacher who liked giving his students long-winded problems and then resting, or on some occasions sleeping, in class. On his first day with Gauss, who was 8 years old, the teacher began, as usual, by telling the students to find the sum of all the numbers from 1 to 100. The teacher barely had a chance to sit down before Gauss raised his hand and said "5050." The dumbfounded teacher, who believed Gauss must have heard the problem before and memorized the answer, asked Gauss to explain how he had solved the problem. He replied: "The numbers 1, 2, 3, 4, 5, and so on to 100 can be paired as 1 and 100, 2 and 99, 3 and 98, and so on. Since each pair has a sum of 101, and there are 50 pairs, the total is 5050."

1

Outline and Objectives

VIGNETTE: THERE'S NO SLEEPING WHEN HE'S AROUND!

INTRODUCTION

1–1 WHOLE NUMBERS AND COUNTING

Objective 1: Describe the origination of the decimal number system.

1–1–1 Base or Radix of a Number System

Objective 2: Explain the basic process of counting in decimal, and the following terms.
  a. Base or radix
  b. Positional weight
  c. Reset and carry

1–1–2 Positional Weight

1–1–3 Reset and Carry

1–2 REPRESENTING FRACTIONS

Objective 3: Describe what a fraction is, and name its elements.

Objective 4: Demonstrate how fractions are represented.

1–3 ADDING FRACTIONS

Objective 5: Describe how fractions are added.

1–3–1 Method 1 for Finding a Common Denominator

Objective 6: Show two methods used for calculating the lowest common denominator.

1–3–2 Method 2 for Finding a Common Denominator

1–3–3 Improper Fractions and Mixed Fractions

Objective 7: Define a(n):
  a. Proper fraction
  b. Improper fraction
  c. Mixed fraction

1–3–4 Adding Mixed Numbers

Objective 8: Show how mixed fractions are added.

1–3–5 Reducing Fractions

Objective 9: Describe how to reduce fractions.

1–4 SUBTRACTING FRACTIONS

Objective 10: Show how to subtract proper and mixed fractions.

1–4–1 Subtracting Mixed Fractions

1–4–2 Borrowing

1–5 MULTIPLYING FRACTIONS

Objective 11: Show how to multiply proper and mixed fractions.

1–5–1 Multiplying Mixed Fractions

1–6 DIVIDING FRACTIONS

Objective 12: Show how to divide proper and mixed fractions.

1–6–1 Dividing Mixed Fractions

1–7 CANCELING FRACTIONS

Objective 13: Demonstrate how to use cancellation to simplify fraction multiplication or division.

Introduction

Since this is your first chapter, let me try to put you in the right frame of mind before you begin. Most people have trouble with mathematics because they never mastered the basics. As you proceed through this chapter and the succeeding chapters, it is imperative that you study every section, example, self-test evaluation point, and end-of-chapter question. If you cannot understand a particular section or example, go back and review the material that led up to the problem and make sure that you fully understand all the basics before you continue. Since each chapter builds on previous chapters, you may find that you need to return to an earlier chapter to refresh your understanding before moving on with the current chapter. This process of moving forward and then backtracking to refresh your understanding is very necessary and helps to engrave the basics of mathematics in your mind. Try never to skip a section or chapter because you feel that you already have a good understanding of that material. If it is a basic topic that you have no problem with, read it anyway to refresh your understanding about the steps involved and the terminology because these may be used in a more complex operation in a later chapter.

1–1 WHOLE NUMBERS AND COUNTING

Thousands of years ago, people used their fingers and thumbs to keep track of objects, and so it is not surprising that our number system has 10 numerals or digits. In fact, the word "digit" means "finger," and the decimal numbers system (from the Latin word *decimus*, which means ten) uses the following digits:

$$0, 1, 2, 3, 4, 5, 6, 7, 8, \text{ and } 9$$

Counting was probably the first application for a number system such as decimal because people needed some way of keeping track of things. *Decimal is based on the number 10.* Let us first examine what is meant by the *base* of a number system.

1–1–1 *Base or Radix of a Number System*

Base or Radix
A number equal to the number of units in a given number system. The decimal system uses a base of 10.

Subscript
A distinguishing letter or numeral written immediately below and to the right or left of another character.

The key element that distinguishes one number system from another is the **base** *or* **radix** *of the number system*. The base describes the number of digits that are used. The decimal number system has a base of 10, which means that it makes use of 10 digits (0 through 9) to represent the value of a quantity. A **subscript,** which is a smaller number written to the right or left of and below the main number, is sometimes included to indicate the number's base. For example, all the numbers following have the subscript "$_{10}$" to indicate that they are base 10, or decimal, numbers.

If a subscript is not present, the number is assumed to be a base 10, or decimal, number. For example, the following are all decimal numbers, even though they do not include the subscript 10.

$$23 \quad 101 \quad 3{,}634{,}987 \quad 47$$

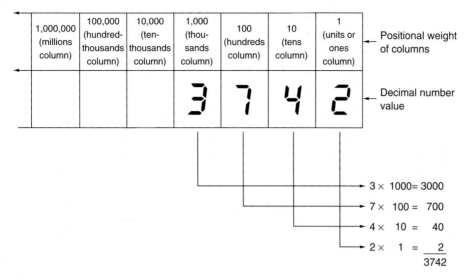

FIGURE 1–1 Positional Weight of Each Decimal Column.

1-1-2 *Positional Weight*

The position of each digit of a decimal number determines the weight of that digit. For example, a 2 by itself indicates only a value of two, whereas a 2 to the left of three zeros (2000) indicates a value of two thousand. The position of a decimal digit therefore determines its weight, or value, with digits to the left carrying the greater weight. For example, the decimal number 3742 contains three thousands (3×1000), seven hundreds (7×100), four tens (4×10), and two ones (2×1). Figure 1–1 shows this decimal number with the positional weight of each decimal column.

You may have noticed that the weight of each decimal (base 10) column increases by a factor of 10 as you move to the left. For example, the units column has a weight of 1. The next column to the left of the units column (tens column) has a weight that is ten times ($\times 10$) larger than the units column—$10 \times$ one $= 10$. The next column to the left of the tens column (hundreds column) has a weight ten times ($\times 10$) larger than the tens column—$10 \times$ ten $= 100$. Therefore, each column in the decimal number system increases by ten times ($\times 10$) as you move to the left.

1-1-3 *Reset and Carry*

Figure 1–2 shows the reset and carry action that occurs when we count in decimal. *This* **reset and carry action** *occurs whenever a column reaches the highest decimal digit of 9*. For example, when we begin to count from zero (0), the units column continues to advance by one or **increment** until it reaches 9. An increment beyond this point causes the units column to be reset to zero and a one to be carried into the tens column, producing 10 (ten). As this count continues, the units column will again repeat the cycle going from 0 to 9, which with the one in the tens column will be a count of 10 (ten) to 19 (nineteen). An increase by one beyond 19 will again cause the units column to reset to zero and carry a one into the tens column, resulting in a count of 20 (twenty). The units column will continue to cycle from 0 to 9 and reset and carry into the tens column until the tens column reaches the highest decimal digit, 9. As you can see in Figure 1–2, an advance by one beyond a count of 99 will cause the units column to reset and carry one into the tens column, and because the tens column is at its maximum, 9, it will also reset to zero and carry one into the hundreds column. This "reset to zero and carry one into the next left column action" will continue to occur in a right-to-left motion whenever a column reaches the maximum digit 9. In other words, the units column will reset and carry into the tens column, the tens column will reset and carry into the hundreds column, the hundreds column will reset and carry into the thousands column, and so on.

FIGURE 1–2 The Reset and Carry Action That Occurs When Counting in Decimal.

Reset and Carry Action
An operation that occurs when counting, in which a digit resets after advancing beyond its maximum count and carries a one into the next higher order column.

Increment
The action or process of increasing a quantity or value by one.

All of the counting numbers discussed so far are in fact whole numbers, since a whole number, by definition, is a single digit or multiple digit complete number such as 7,236, or 15,250.

Some whole numbers are followed by a fraction. For example, with the number *twelve and one-half* $\left(12\frac{1}{2}\right)$, the number *twelve* is a whole number while the *one-half* is a fractional part or piece of the whole. Fractions will be discussed in the next section.

SELF-TEST EVALUATION POINT FOR SECTION 1–1

Now that you have completed this section, you should be able to:

■ **Objective 1.** Describe the origination of the decimal number system.

■ **Objective 2.** Explain the basic process of counting in decimal, and the terms *base or radix*, *positional weight, and* reset and carry.

Use the following questions to test your understanding of Section 1–1.

1. The decimal number system has a base or radix of _____.

2. Indicate the positional weight of each of the decimal digits in the following number: 35,729.

3. What is the reset and carry action?

4. Which columns will reset and carry when the number 499,997 is advanced or incremented by 5? What will be the final count?

Use the following technical trade questions to test your understanding of practical applications of Section 1–1.

1. **Electrical** A spool of wire is listed as weighing 147 pounds. If an additional 9 pounds of wire was added to the spool, what would be the total weight?

2. **Machining** What is the distance between holes in the pictured machined part?
   a. *x* and *y*?
   b. *y* and *z*?

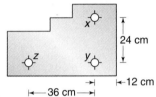

3. **Automotive** To square a damaged car frame, an autobody worker measured diagonals between the two front cross-members and rear cross-members. The front had to be adjusted 4 cm to be equal, while the rear had to be adjusted half that amount. How much was the rear adjusted?

4. **Plumbing** If 7 cm is cut off a 20 cm pipe, and then a 2 cm coupling is welded onto one end, what is the new length of the pipe and coupling?

5. **Welding** A welder determines that 7 cubic feet of acetylene gas is needed to make a supporting frame for a barbecue. How many cu. ft. of gas is needed to make three?

6. **Construction** After completing the construction of a garden shed, $8 of tax needs to be added to the total billed amount of $107. What is the total cost?

1–2 REPRESENTING FRACTIONS

Fraction
A numerical representation indicating the quotient of two numbers; a piece, fragment, or portion.

Half
Either of two equal parts into which a thing is divisible.

Quarter
One of four equal parts into which something is divisible; a fourth part.

Fourth
See *quarter*.

Another word for **fraction** is "piece"; therefore, if you have a fraction of something, you have a piece of that something. Having only a fraction or piece of something means that the object has been broken up into pieces, and therefore a fraction is something less than the whole thing. For example, if you wanted half of a pizza, a whole pizza would be cut into two pieces and you would be given one of the halves—a **half** is therefore a fraction. If, in another instance, you wanted only a quarter of an apple pie, a whole apple pie would be cut into four pieces and you would be given one of the quarters—a **quarter** is therefore also a fraction. A quarter (or a **fourth**) is a smaller fraction than a half, and therefore we can say that *the more cuts you make, the smaller the fraction.*

Half and *quarter* (or *fourth*) are names that describe a fraction; however, some other system, sign, or symbol is needed to represent or show the fraction. This system will have to show two things:

1. The number of pieces into which the entire object was cut or divided up
2. The number of pieces we have or are concerned with

■ EXAMPLE

Write down the proper or written fraction for one-quarter or one-fourth, and label its elements.

■ *Solution:*

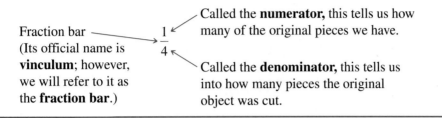

| **Numerator** |
| The part of a fraction that is above the line and signifies the number of parts of the denominator taken. |

| **Denominator** |
| The part of a fraction below the line signifying division that functions as the divisor of the numerator. |

| **Vinculum or Fraction Bar** |
| A straight horizontal line placed between two or more numbers of a mathematical expression. |

The preceding example fraction represents an amount or quantity. In this example it represents "one-quarter" or "one-fourth" and indicates that the object was cut or divided into four pieces and that we have one of the four pieces.

■ EXAMPLE

First, draw a picture of a pizza cut into six pieces,

   a. then show one of the six pieces separately and label it with its fraction
   b. then show three of the six pieces separately and label them with their fraction.

■ *Solution:*

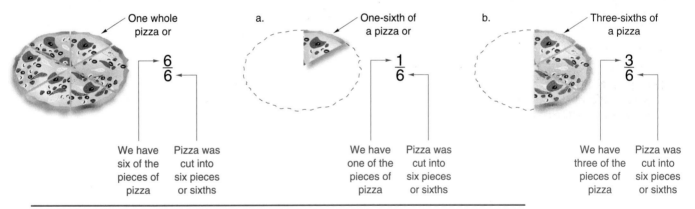

In all technical fields you will have to interpret scales, such as the divisions on the straightedge shown next.

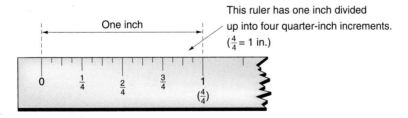

Other rulers might have each inch broken up into halves (there would be two halves or $\frac{2}{2}$ in an inch), eighths (there would be eight eighths or $\frac{8}{8}$ in an inch), sixteenths (there would be sixteen sixteenths or $\frac{16}{16}$ in an inch), and so on.

SECTION 1–2 / REPRESENTING FRACTIONS

TECHNICAL TRADE APPLICATION: CONSTRUCTION

Use the ruler to measure the length of the screws shown in the following figure.

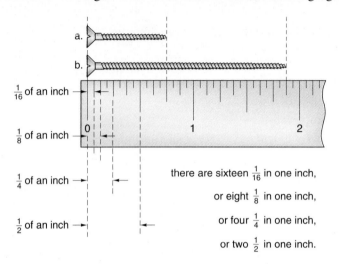

■ *Solution:*

a. The smaller screw measures three-fourths of an inch ($\frac{3}{4}$ in.), or six-eighths of an inch ($\frac{6}{8}$ in.), or twelve-sixteenths ($\frac{12}{16}$ in.) of an inch.

b. The second screw is greater than 1 inch long but less than 2 inches long, which means that it is 1 inch plus a fraction of an inch long. Since its tip falls alongside an $\frac{1}{8}$ inch mark on the ruler, we will use the $\frac{1}{8}$ inch scale. The longer screw measures one and seven-eighths of an inch ($1\frac{7}{8}$ in.) long. We could also say that screw b measures $\frac{15}{8}$ of an inch, since the $\frac{8}{8}$ in the first inch, plus the $\frac{7}{8}$ in the second inch, equals a total of $\frac{15}{8}$. When we use the sixteenths ruler marks, screw b measures $1\frac{14}{16}$ or $\frac{30}{16}$ of an inch long.

SELF-TEST EVALUATION POINT FOR SECTION 1–2

Now that you have completed this section, you should be able to:

■ *Objective 3.* Describe what a fraction is, and name its elements.

■ *Objective 4.* Demonstrate how fractions are represented.

Use the following questions to test your understanding of Section 1–2.

1. What fraction of each of the circular objects is shown shaded?
2. What fraction of each of the circular objects is shown unshaded?

a.

b.

c.

Use the following technical trade questions to test your understanding of practical applications of Section 1–2.

1. **Carpentry** If a carpenter divides a 6-foot plank of wood into three parts, each part would be _____ of the whole, and so equal to _____.

   a. $\frac{1}{3}$ (one-third), 3 feet
   b. $\frac{1}{2}$ (one-half), 2 feet
   c. $\frac{1}{4}$ (one-fourth), 4 feet
   d. $\frac{1}{3}$ (one-third), 2 feet

2. **Automotive** An alternator is mounted to its bracket using one-half inch bolts. Which of the following would be equivalent?

   a. $\frac{2}{4}$
   b. $\frac{3}{6}$
   c. $\frac{4}{8}$
   d. All of the above.

3. **Construction** As shown below, a roof's pitch or steepness is defined as a fraction:

$$\text{Pitch} = \frac{\text{Rise}}{\text{Run}} \quad \begin{array}{l} \text{Roof's rise—Increase in height} \\ \text{Roof's run—Horizontal distance} \end{array}$$

What fraction would define the roof shown in the figure?

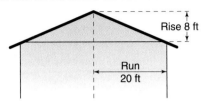

4. **Electrical** Connecting an ammeter to a robotic arm, an electrician monitors the current drawn at different stages of the manufacturing process.

What are the ampere readings shown on the scale, at points a through e?

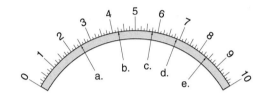

5. **Printing** To print 20,000 newspapers, a printer uses 14 of the 30 rolls of newsprint in the warehouse. Express the number of rolls remaining of the total as a fraction.

6. **Plumbing** Which of the following is the larger pipe size?
   a. $\frac{1}{4}$ in. c. $\frac{7}{8}$ in.
   b. $\frac{3}{4}$ in. d. $\frac{1}{2}$ in.

7. **Landscaping** The instructions on the container of a liquid fertilizer state that 4 ounces of fertilizer should be mixed with 68 ounces of water. What fraction represents the amount of liquid fertilizer to total mixture?

8. **Metalwork** Which sheet metal is thicker?
   a. $\frac{2}{4}$ inch
   b. $\frac{1}{2}$ inch

9. **Machining** If a machinist produces 4 units in 8 hours $\left(\frac{4 \text{ units}}{8 \text{ hours}}\right)$, how long is it taking to manufacture 1 unit?

10. **Administration** The office printer uses one $89.95 printer cartridge for every 5 reams of paper (one ream has 500 sheets of paper and costs $3.97). Express the total paper cost for each cartridge cost as a fraction.

1–3 ADDING FRACTIONS

The *denominator* (number below the fraction bar) determines the fraction type. For example, $\frac{3}{9}$ tells us that the object was divided up into nine parts and that we have three of those nine parts. The number 9 therefore tells us the type of fraction we are dealing with. When fractions are of the same type (have the same denominators), they are added as simply as adding 5 apples and 3 apples.

■ EXAMPLE

Add the following fractions: $\frac{5}{9} + \frac{3}{9}$

■ Solution:

Since we are dealing with two fractions that are of the same type (ninths), the result is simply the sum or total of the two *numerators* (numbers above the fraction bar).

$$\begin{array}{c} \text{Numerators} \rightarrow \\ \text{Denominators} \rightarrow \end{array} \frac{5}{9} + \frac{3}{9} = \frac{8}{9}$$

Notice that the numerators were added, whereas the denominators remained the same. This is because the type of fraction remained the same; only the number of pieces changed. To use an analogy, we could say:

$$\frac{5}{\text{apples}} + \frac{3}{\text{apples}} = \frac{8}{\text{apples}} \begin{array}{l} \leftarrow \text{Sum} \\ \leftarrow \text{Type} \end{array}$$

The next step is to address how we can add different types of fractions. For example, how would we add $\frac{1}{2}$ plus $\frac{1}{4}$? In this instance we have two different types of fractions, and therefore they are as different as apples and oranges.

Using pictures, we know that $\frac{1}{2}$ plus $\frac{1}{4}$ equal $\frac{3}{4}$.

The question is: What mathematical procedure do we follow when the answer is not this obvious? If you study the previous answer, you see that the answer was in fourths. Therefore, what we actually did was convert the one-half into two-fourths ($\frac{1}{2} = \frac{2}{4}$, one-half = two-fourths) and then add the other fourth to get three-fourths.

The written fractions would appear as follows:

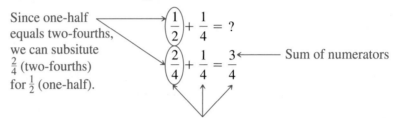

Since one-half equals two-fourths, we can subsitute $\frac{2}{4}$ (two-fourths) for $\frac{1}{2}$ (one-half).

Sum of numerators

Now the fractions are the same type; therefore the denominator answer indicates the same fraction type.

What we have noticed from this example is that to add fractions that are of different types, we need to find a *denominator that is common to both fractions*. There are two methods for finding a **common denominator** so that we can add different types of fractions.

Common Denominator
A common multiple of the denominators of a number of fractions.

1–3–1 *Method 1 for Finding a Common Denominator*

Probably the easiest way to find a common denominator is to multiply one denominator by another. For example, let us assume that we want to add $\frac{1}{2}$ plus $\frac{1}{3}$.

$$\frac{1}{2} + \frac{1}{3} = ?$$

Our common denominator can be obtained by multiplying one denominator by the other; therefore, $2 \times 3 = 6$.

Now that a common denominator has been found, how many sixths $\left(\frac{1}{6}\right)$ are in one-half $\left(\frac{1}{2}\right)$ and one-third $\left(\frac{1}{3}\right)$? This is determined in the following way.

$$\frac{1}{2} + \frac{1}{3} = \frac{?}{6}$$

(One-half plus one-third equals how many sixths?)

$$\frac{1}{2} + \frac{1}{3} = \frac{3}{6}$$

Step a: 2 into 6 = 3.
Step b: 3 × 1 = 3.
Step c: Place result in answer's numerator.

$$\frac{1}{2} + \frac{1}{3} = \frac{3 + 2}{6}$$

Step a: 3 into 6 = 2.
Step b: 2 × 1 = 2.
Step c: Place result in answer's numerator.

$$\frac{1}{2} + \frac{1}{3} = \frac{3 + 2}{6}$$

The last step is to add the two answers for the numerator (3 + 2 = 5).

$$\frac{1}{2} + \frac{1}{3} = \frac{5}{6}$$

■ **EXAMPLE**

Add $\frac{1}{4}$ and $\frac{2}{3}$

■ *Solution:*

Common denominator = denominator × denominator

$$= 4 \times 3$$
$$= 12$$

$$\frac{1}{4} + \frac{2}{3} = \frac{3 + 8}{12} = \frac{11}{12}$$

Steps:
4 into 12 = 3,
3 × 1 = 3,
3 into 12 = 4,
4 × 2 = 8,
3 + 8 = 11.

1–3–2 *Method 2 for Finding a Common Denominator*

The problem with method 1 is that it often yields a common denominator that is larger than necessary. For example, what is the lowest-value common denominator for the following three-fraction addition?

$$\frac{3}{9} + \frac{1}{4} = \frac{1}{6} = \frac{}{?}$$

Using method 1, we would obtain the answer

$$9 \times 4 \times 6 = 216$$

Even though 216 will work as a common denominator, it is a very large number and therefore the fraction addition will be cumbersome.

The steps in the following method will always yield the lowest **common denominator.** Using the previous example, we first place all three denominators in an upside-down division box, as shown.

$$\underline{9 \quad 4 \quad 6}$$

The next step is to begin with 2 and see if it will divide exactly into any of the three denominators. Once 2 can no longer be used as a divisor, increase to 3, and then 4, and then 5, and so on. Let us follow through these steps in more detail.

Lowest Common Denominator
The lowest-value common denominator.

2 does not go into 9 exactly.
2 goes into 4, giving 2,
and 2 goes into 6, giving 3.
Using 2 again as a divisor,
we can further reduce because 2
goes into 2, giving 1.
Because 2 will no longer reduce
any of the denominators, we increase
the divisor to 3. 3 goes into
9, giving 3, and into 3, giving 1.
Using 3 again as a divisor, we
can reduce because 3 goes into
3, giving 1.

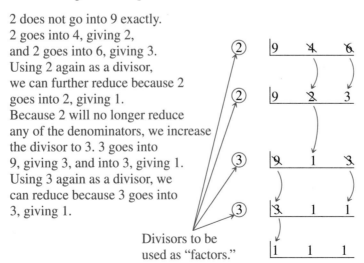

Divisors to be used as "factors."

Looking at the previous process, you can see that the point is to keep dividing the denominators until they are all reduced to 1. Because no number is changed when it is multiplied or divided by 1 (for example, 22 × 1 = 22, 115 × 1 = 115, 17 ÷ 1 = 17, 63 ÷ 1 = 63), we can ignore the remaining 1s.

The divisors used in this process are called *factors*. A **factor** *is any number that is multiplied by another number and contributes to the result.* For example, consider 3 × 8. In this multiplication example, 3 is a factor and 8 is a factor that will contribute to the result 24. Let us now take another example: What factors could contribute to a result of 12? In this case we must determine which small numbers, when multiplied together, will produce 12. The answer is

$$2 \times 6 = 12 \quad \text{or} \quad 3 \times 4 = 12$$

Are these, however, the smallest numbers that when multiplied together will produce 12? The answer is no because the 6 can be broken down into 2 × 3, and the 4 can be broken down into 2 × 2. Therefore, 2 × 3 × 2 or 3 × 2 × 2 shows that 12 has the three factors 2, 2, and 3.

Some numbers cannot be reduced or broken down into factors. For example, there are no smaller numbers that when multiplied together will produce 5. This number is called a **prime number,** and there are many, such as 3, 5, 7, 11, 13, and 17.

What we were doing, therefore, in the previous three-fraction addition example was extracting the factors from the denominators.

> **Factor**
> Any of the numbers or symbols in mathematics that when multiplied together form a product.

> **Prime Number**
> Any integer other than 0 or 1 that is not divisible without remainder.

These four factors, which were the divisors, can then be used to find the lowest common denominator.

$$\begin{array}{r|rrr} 2 & 9 & 4 & 6 \\ 2 & 9 & 2 & 3 \\ 3 & 9 & 1 & 3 \\ 3 & 3 & 1 & 1 \end{array} \quad (\text{Example: } \frac{3}{9} + \frac{1}{4} + \frac{1}{6} = \frac{?}{?})$$

$$\underbrace{2 \times 2 \times 3 \times 3}_{\text{Denominator factors}} = \underbrace{36}_{\text{Lowest common denominator}}$$

> **Factoring**
> Resolving into factors; a reducing process that extracts the common factors.

Now that we have completed the *factoring* step and determined the lowest common denominator, we can proceed with the example, which was to add $\frac{3}{9} + \frac{1}{4} + \frac{1}{6}$:

$$\frac{3}{9} + \frac{1}{4} + \frac{1}{6} = \frac{}{36} \longleftarrow \text{Lowest common denominator}$$

$$\left(\frac{3}{9}\right) + \frac{1}{4} + \frac{1}{6} = \frac{12 +}{36} \qquad \textit{Steps:} \quad \begin{array}{l} 9 \text{ into } 36 = 4, \\ 4 \times 3 = 12. \end{array}$$

$$\frac{3}{9} + \left(\frac{1}{4}\right) + \frac{1}{6} = \frac{12 + 9 +}{36} \qquad \begin{array}{l} 4 \text{ into } 36 = 9, \\ 9 \times 1 = 9. \end{array}$$

$$\frac{3}{9} + \frac{1}{4} + \left(\frac{1}{6}\right) = \frac{12 + 9 + 6}{36} \qquad \begin{array}{l} 6 \text{ into } 36 = 6, \\ 6 \times 1 = 6. \end{array}$$

$$= \frac{12 + 9 + 6}{36} = \frac{27}{36}$$

■ **EXAMPLE**

Add $\frac{1}{4} + \frac{3}{5} + \frac{1}{8}$

CHAPTER 1 / WHOLE NUMBERS AND FRACTIONS

■ *Solution:*

The first step is to determine the factors in each denominator.

Divisors or factors 3 and 4 were ignored because 5 was a prime factor.

$$
\begin{array}{r|ccc}
2 & 4 & 5 & 8 \\
2 & 2 & 5 & 4 \\
2 & 1 & 5 & 2 \\
5 & 1 & 5 & 1 \\
  & 1 & 1 & 1
\end{array}
$$

$2 \times 2 \times 2 \times 5 = 40$ (Lowest common denominator)

$$\frac{1}{4} + \frac{3}{5} + \frac{1}{8} = \frac{10 + 24 + 5}{40} = \frac{39}{40}$$

1–3–3 *Improper Fractions and Mixed Numbers*

All the fractions discussed so far have a numerator that is smaller than the denominator. These are called **proper fractions.** For example:

Proper fractions: $\dfrac{1}{2} \quad \dfrac{3}{4} \quad \dfrac{5}{9} \quad \dfrac{7}{32} \quad \dfrac{15}{27}$ ← Smaller numerator
← Larger denominator

Proper Fraction
A fraction in which the numerator is less or of lower degree than the denominator.

On some occasions, an addition of two or more proper fractions will produce a fraction with a numerator that is larger than the denominator. To explain this, let us use an example.

■ **EXAMPLE**

Add $\dfrac{3}{4} + \dfrac{2}{4}$

■ *Solution:*

$$\frac{3}{4} + \frac{2}{4} = \frac{3+2}{4} = \frac{5}{4} \leftarrow \text{Numerator is larger than denominator.}$$

To analyze this answer, $\frac{5}{4}$, we would say that the denominator 4 indicates that the original whole unit was divided into four pieces (quarters) and the numerator indicates that we have five of these pieces. Therefore, with four of the quarters (or fourths) we have one whole unit, leaving one extra fourth.

$$\frac{5}{4} = 1\frac{1}{4}$$

five-fourths = one whole unit and one-fourth

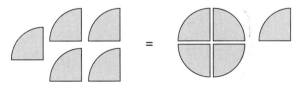

Improper Fraction
A fraction whose numerator is equal to or larger than the denominator.

Mixed Number
A number composed of an integer and a fraction.

A fraction with a numerator that is larger than the denominator (for example, $\frac{5}{4}$) is called an **improper fraction** and needs to be converted to a **mixed number.** By definition, a mixed fraction or mixed number is a whole number and a fraction (for example, $1\frac{1}{4}$). Since the fraction bar actually indicates the arithmetic operation of division (for example, $\frac{5}{4}$ indicates 5 divided by 4), all we do is perform this division to convert an improper fraction to a mixed number. For example, 4 goes into 5 once with 1 remaining:

$$\frac{5}{4} = 1\frac{1}{4}$$

four into five = one, with one-fourth remaining

■ EXAMPLE

Add $\frac{2}{3} + \frac{4}{6}$

■ Solution:

$$\frac{2}{3} + \frac{4}{6} = \frac{4+4}{6} = \frac{8}{6}$$

Lowest common denominator:

$$\begin{array}{ccc} 2 & 3 & 6 \\ 3 & 3 & 3 \end{array}$$
$$\rightarrow 2 \times 3 = 6$$

Because $\frac{8}{6}$ is an improper fraction, the next step is to convert it to a mixed fraction.

$$\frac{8}{6} = 1\frac{2}{6}$$

six into eight = one, with two-sixths remaining

1-3-4 Adding Mixed Numbers

Now that we understand what a mixed number is, let us see how we would add two mixed fractions.

■ EXAMPLE

Add $5\frac{1}{4} + 6\frac{3}{8}$

■ Solution:

The easiest way to add mixed numbers is to deal with the whole numbers and fractions separately.

Step 1: Separate whole number and fractions.

Step 2: Add fractions.

Step 3: Add whole numbers.

Step 4: Combine whole-number result and fraction result.

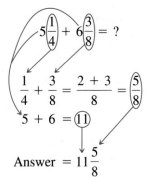

$5\frac{1}{4} + 6\frac{3}{8} = ?$

$\frac{1}{4} + \frac{3}{8} = \frac{2+3}{8} = \frac{5}{8}$

$5 + 6 = 11$

Answer $= 11\frac{5}{8}$

TECHNICAL TRADE APPLICATION: CARPENTRY

If two wood screws measure $6\frac{3}{8}$ inches and $2\frac{1}{4}$ inches, what is the total length of both?

■ *Solution:*

$$6\frac{3}{8} + 2\frac{1}{4} = 8\frac{3+2}{8} = 8\frac{5}{8} \text{ inches}$$

■ EXAMPLE

Add $4\frac{15}{20} + 5\frac{4}{5}$

■ *Solution:*

$$4\frac{15}{20} + 5\frac{4}{5} = 9\frac{15+16}{20} = 9\frac{31}{20}$$

Because the fraction $\frac{31}{20}$ is improper, we must convert it by dividing 20 into 31.

$$9\frac{31}{20} = 9 + \left(1\frac{11}{20}\right) \longleftarrow 31 \div 20 = 1\frac{11}{20}$$

$$= 10\frac{11}{20} \longleftarrow \begin{array}{l} \text{9 whole objects plus} \\ \text{1 whole object} = 10 \end{array}$$

1-3-5 *Reducing Fractions*

To obtain half of a pizza politely, you could ask for two small quarters because two-quarters equals one-half.

$$\frac{2}{4} = \frac{1}{2}$$

Both fractions therefore indicate the same part of a whole object. In most instances when we have two fractions that describe the same portion, we would use the fraction that has the smaller denominator. For example, the denominator 2 in $\frac{1}{2}$ is smaller than the denominator 4 in $\frac{2}{4}$, and therefore we would use $\frac{1}{2}$ to represent the fraction. The next question is: How do we **reduce** a fraction so that it has the lowest possible denominator? The answer is that we use an algebraic rule that states: *If the same number is divided into both the numerator and denominator of a fraction, the value of the fraction remains the same, only reduced.*

For example, let us divide 2 into both the numerator and denominator of $\frac{2}{4}$:

$$\frac{2 \div 2}{4 \div 2} = \frac{1}{2}$$

Reducing
The process of changing to an equivalent but more fundamental expression.

To reduce a fraction we must therefore find the largest number that will divide evenly into both its numerator and denominator. This is best achieved by starting with 2, then trying 3, then 4, then 5, and so on. To demonstrate this, let us do the following example.

■ EXAMPLE

Reduce the following fractions if possible.

a. $\dfrac{3}{6}$ b. $\dfrac{4}{16}$ c. $\dfrac{5}{45}$ d. $\dfrac{4}{9}$ e. $\dfrac{18}{90}$

■ Solution:

a. $\left(\dfrac{3}{6}\right) \begin{array}{c} \div 3 \\ \div 3 \end{array} = \dfrac{1}{2}$ (three-sixths = one-half)

   — Will both divide by 2? No. Will both divide by 3? Yes.

b. $\left(\dfrac{4}{16}\right) \begin{array}{c} \div 4 \\ \div 4 \end{array} = \dfrac{1}{4}$ (four-sixteenths = one-fourth)

   — Will both divide by 2? Yes. Will both divide by 3? No. Will both divide by 4? Yes. We will use 4 because dividing by a larger number will give us a smaller denominator in fewer steps. If we had used 2, we would have ended up with the same answer but would have had to reduce in two steps:

$$\left(\dfrac{4}{16}\right) \begin{array}{c} \div 2 \\ \div 2 \end{array} = \left(\dfrac{2}{8}\right) \begin{array}{c} \div 2 \\ \div 2 \end{array} = \dfrac{1}{4}$$

c. $\left(\dfrac{5}{45}\right) \begin{array}{c} \div 5 \\ \div 5 \end{array} = \dfrac{1}{9}$

   — Will both divide by 2? No. Will both divide by 3? No. Will both divide by 4? No. Will both divide by 5? Yes.

d. $\left(\dfrac{4}{9}\right)$

   — Will both divide by 2? No. Will both divide by 3? No. Will both divide by 4? No. This fraction cannot be reduced any further.

e. $\left(\dfrac{18}{90}\right) \begin{array}{c} \div 18 \\ \div 18 \end{array} = \dfrac{1}{5}$

   — Divide by 2? Yes. Divide by 3? Yes. Divide by 4? No. Divide by 5? No. Divide by 6? Yes. Divide by 7? No. Divide by 8? No. Divide by 9? Yes. … Divide by 18? Yes.

SELF-TEST EVALUATION POINT FOR SECTION 1–3

Now that you have completed this section, you should be able to:

■ **Objective 5.** *Describe how fractions are added.*

■ **Objective 6.** *Show two methods used for calculating the lowest common denominator.*

■ **Objective 7.** *Define a proper fraction, an improper fraction, and a mixed fraction.*

■ **Objective 8.** *Show how mixed fractions are added.*

■ **Objective 9.** *Describe how to reduce fractions.*

CHAPTER 1 / WHOLE NUMBERS AND FRACTIONS

Use the following questions to test your understanding of Section 1–3.

Add the following fractions. Convert to proper fractions and reduce if necessary.

1. $\dfrac{4}{6} + \dfrac{1}{6} = ?$
2. $\dfrac{3}{64} + \dfrac{2}{64} + \dfrac{25}{64} = ?$
3. $\dfrac{3}{9} + \dfrac{6}{18} = ?$
4. $\dfrac{1}{3} + \dfrac{1}{4} + \dfrac{3}{15} = ?$
5. $\dfrac{3}{4} + \dfrac{4}{5} = ?$
6. $\dfrac{9}{12} + \dfrac{4}{24} + \dfrac{3}{4} = ?$
7. $\dfrac{7}{9} + \dfrac{5}{9} + \dfrac{4}{18} = ?$

Convert to proper fractions and/or reduce if necessary.

8. $\dfrac{15}{6} = ?$
9. $\dfrac{4}{24} = ?$
10. $\dfrac{25}{100} = ?$

Use the following technical trade questions to test your understanding of practical applications of Section 1–3.

1. **Electrical** What minimum size guide hole should be made to accomodate a $\dfrac{3}{4}$ inch and $1\dfrac{1}{2}$ inch electrical pipe conduit?
2. **Automotive** An automotive technician replaces $2\dfrac{1}{2}$ quarts of oil, $4\dfrac{1}{3}$ quarts of coolant, $\dfrac{1}{4}$ quart of brake fluid, and $3\dfrac{1}{6}$ quarts of washer fluid for the 15,000-mile routine service. On the maintenance sheet, what should be written alongside "total number of quarts used"?
3. **Machining** What is the total length of the machine part shown in the figure?

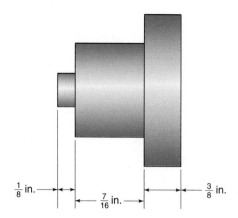

4. **Printing** Newspapers sell their advertising space by the column-inch (c.i.). If an advertiser buys the following over a month, what would be the total?

$9\dfrac{3}{4}$ c.i., $8\dfrac{1}{3}$ c.i., $16\dfrac{1}{8}$ c.i., $8\dfrac{1}{2}$ c.i.

5. **Plumbing** Calculate the total length of the water pipe shown.

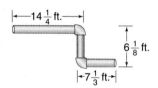

6. **Manufacturing** What is the total length of the metal casting shown?

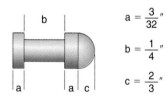

1–4 SUBTRACTING FRACTIONS

To subtract one fraction from another, we follow the same steps as in addition except that we subtract the numerators. Let us examine a few examples, starting with a subtraction that involves two fractions with the same denominator.

■ **EXAMPLE**

What is $\dfrac{9}{16} - \dfrac{4}{16}$?

■ **Solution:**

$$\dfrac{9}{16} - \dfrac{4}{16} = \dfrac{9-4}{16} = \dfrac{5}{16}$$

If the two fractions involved in the subtraction are of different types, we will need to determine the lowest common denominator so that we are subtracting one type of fraction from the same type of fraction.

■ **EXAMPLE**

What is the result of $\dfrac{6}{10} - \dfrac{4}{20}$?

■ *Solution:*

$$\dfrac{6}{10} - \dfrac{4}{20} = \dfrac{12 - 4}{20} = \dfrac{8}{20}$$
$$= \dfrac{8 \div 4}{20 \div 4} = \dfrac{2}{5}$$

Steps:
Find lowest common denominator (20).
10 into 20 = 2,
2 × 6 = 12.
20 into 20 = 1,
1 × 4 = 4.
12 − 4 = 8.
Reduce result.

1-4-1 Subtracting Mixed Fractions

As with addition, the way to subtract one mixed fraction from another is to deal with the whole numbers and fractions separately.

■ **EXAMPLE**

Subtract $2\dfrac{1}{3}$ from $5\dfrac{4}{5}$

■ *Solution:*

$$5\dfrac{4}{5} - 2\dfrac{1}{3} = 3\dfrac{12 - 5}{15}$$
$$= 3\dfrac{7}{15}$$

Steps:
Subtract whole numbers: 5 − 2 = 3.
Determine the lowest common denominator.

3 | 5   3
5 | 5   1
↳ 3 × 5 = 15

5 into 15 = 3, 3 × 4 = 12.
3 into 15 = 5, 5 × 1 = 5.
12 − 5 = 7.

1-4-2 Borrowing

As is normal in subtraction, we sometimes have problems in which we need to subtract a large number from a small number. In these cases we will need to **borrow**, as shown in the following example.

Borrowing
The process of taking from the next higher order digit so that a subtraction can take place.

■ **EXAMPLE**

What is $6\dfrac{1}{3} - 2\dfrac{2}{3}$?

■ **Solution:**

Borrow 1 from 6 and convert it to $\frac{3}{3}$.

$$6\frac{1}{3} - 2\frac{2}{3} = 4\frac{1-2}{3}?$$

$$6\frac{1}{3} = 5 + \left(\frac{3}{3}\right) + \frac{1}{3} = 5\frac{4}{3}$$

$$5\frac{4}{3} - 2\frac{2}{3} = 3\frac{4-2}{3} = 3\frac{2}{3}$$

Steps:
Subtract whole numbers; $6 - 2 = 4$.
Because fractions have the same denominator, subtract numerators.
You cannot subtract 2 from 1.
Borrow whole number from 6, and convert it to a fraction $\left(1 = \frac{3}{3}\right)$.
Add 3 pieces to 1 piece to get 4 pieces, or 4 thirds.
Perform subtraction.

■ **TECHNICAL TRADE APPLICATION: BANKING**

The interest rate on a car loan has been reduced $2\frac{1}{4}$ percent from $6\frac{3}{16}$ percent. What is the new interest rate?

■ **Solution:**

$$6\frac{3}{16} - 2\frac{1}{4} = 4\frac{3-4}{16} \longleftarrow \text{You cannot subtract 4 from 3.}$$

$$5 + \frac{16}{16} + \frac{3}{16} - 2\frac{1}{4} =$$

$$5\frac{19}{16} - 2\frac{1}{4} = 3\frac{19-4}{16} = 3\frac{15}{16}$$

SELF-TEST EVALUATION POINT FOR SECTION 1–4

Now that you have completed this section, you should be able to:

■ **Objective 10.** *Show how to subtract proper and mixed fractions.*

Use the following questions to test your understanding of Section 1–4.

1. $\frac{4}{32} - \frac{2}{32} = ?$
2. $\frac{4}{5} - \frac{3}{10} = ?$
3. $\frac{6}{12} - \frac{4}{16} = ?$
4. $15\frac{1}{3} - 2\frac{4}{12} = ?$
5. $5\frac{15}{36} - 2\frac{31}{36} = ?$

Use the following technical trade questions to test your understanding of practical applications of Section 1–4.

1. **Carpentry** What is the difference in height between the two cabinets shown?

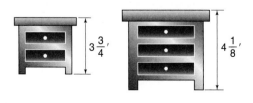

2. **Drafting** Determine the missing dimension (x).

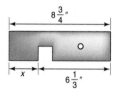

3. **Surveying** What is the difference in height between the two drains shown?

4. **Electrical** What is the power rating difference between a $\frac{1}{4}$ watt resistor and a $2\frac{1}{2}$ watt resistor?

5. **Machining** The diameter of a steel cylinder is reduced $\frac{14}{1000}$ inch from its original $\frac{750}{1000}$ inch. What is the cylinder's new diameter after being machined?

6. **Plumbing** The inside diameter of a steel pipe measures $\frac{5}{8}$ inch $\left(\text{i.d.} = \frac{5}{8}''\right)$ while the outside diameter measures $\frac{12}{16}$ inch $\left(\text{o.d.} = \frac{12}{16}''\right)$. What is the wall thickness of the pipe?

1–5 MULTIPLYING FRACTIONS

One of the best ways to explain something new is to use an example that you can imagine easily.

■ **EXAMPLE**

What is half of a half?

■ *Solution:*

$\frac{1}{2}$ of ◗ = ◢

half of a half = a fourth

$$\frac{1}{2} \times \frac{1}{2} = \frac{1}{4}$$

It is not surprising that the product (multiplication result) of two fractions has a denominator that is smaller than the denominators of the two fractions, because multiplication is calculating the fractional part of a fraction. Let us look at another example.

■ **EXAMPLE**

What is half of a fourth?

■ *Solution:*

$\frac{1}{2}$ of ◢ = ◿

half of one-fourth = one-eighth

$$\frac{1}{2} \times \frac{1}{4} = \frac{1}{8}$$

From the last two examples you can probably see that to obtain the product of two fractions, simply multiply the numerators and multiply the denominators, as shown.

$$\frac{1}{2} \times \frac{1}{2} = \frac{1 \times 1}{2 \times 2} = \frac{1}{4} \quad \text{(one-half of one-half equals one-fourth)}$$

$$\frac{1}{2} \times \frac{1}{4} = \frac{1 \times 1}{2 \times 4} = \frac{1}{8} \quad \text{(one-half of one-fourth equals one-eighth)}$$

■ **EXAMPLE**

What is $\frac{3}{4}$ of $\frac{3}{5}$?

■ *Solution:*

$$\frac{3}{4} \times \frac{3}{5} = \frac{3 \times 3}{4 \times 5} = \frac{9}{20}$$

1–5–1 *Multiplying Mixed Fractions*

When adding or subtracting mixed fractions, we deal with the whole numbers and fractions separately. When multiplying mixed fractions, we must first convert the mixed fractions to improper fractions and then perform the multiplication. For example, what is half of one and one-third?

$$\frac{1}{2} \times 1\frac{1}{3} = ?$$

If we convert $1\frac{1}{3}$ to an improper fraction, we will be dealing only with thirds. Because there are three thirds in one whole $\left(\frac{3}{3} = 1\right)$, these three thirds plus the additional one-third piece equals four-thirds.

$$1\frac{1}{3} = \frac{3}{3} + \frac{1}{3} = \frac{4}{3}$$

Now that the mixed fraction is an improper fraction, we can complete the original problem, which is: What is half of one and one-third, or four-thirds?

$$\frac{1}{2} \times \frac{4}{3} = \frac{4 \div 2}{6 \div 2} = \frac{2}{3}$$

As expected, half of four-thirds equals two-thirds. As you can see, it was necessary to reduce the first result, $\frac{4}{6}$, to $\frac{2}{3}$. This often occurs owing to the larger numbers in the improper fractions.

Before doing another example, let's develop a procedure for converting a mixed fraction to an improper fraction. As an example, we will convert the mixed fraction $2\frac{1}{2}$ to an improper fraction. Because the result will be all halves $\left(2\frac{1}{2} = \frac{?}{2}\right)$, we simply have to determine how many halves are in 2 whole units and then add this value to the additional half. In summary:

$$2\frac{1}{2} = \frac{4}{2} + \frac{1}{2} = \frac{5}{2}$$

Two and one-half = four-halves + one-half = five-halves

To change a mixed fraction to an improper fraction, therefore, you *multiply the whole number by the fraction's denominator and then add the fraction's numerator.*

$$2\frac{1}{2} = \frac{5}{2} \quad \begin{array}{l} \textit{Steps:} \\ \text{Multiply whole number by fraction's} \\ \text{denominator: } 2 \times 2 = 4. \\ \text{Add result to fraction's numerator: } 4 + 1 = 5. \end{array}$$

Now that we understand how to convert a mixed fraction to an improper fraction, we can do a multiplication example involving mixed fractions.

■ EXAMPLE

What is $\frac{4}{9}$ of $5\frac{3}{4}$?

■ Solution:

Note that the term *of* implies multiplication. To determine the result, we must first convert the mixed fraction, $5\frac{3}{4}$, to an improper fraction.

$$5\frac{3}{4} = \frac{23}{4} \quad \begin{array}{l} \textit{Steps:} \\ 5 \times 4 = 20. \\ 20 + 3 = 23. \end{array}$$

The next step is to perform the multiplication using the improper fraction.

$$\frac{4}{9} \times \frac{23}{4} = \frac{92 \div 4}{36 \div 4} = \frac{23}{9}$$

Reduce

Once reduced, the answer may be an improper fraction and therefore will have to be converted back into a mixed fraction. With $\frac{23}{9}$, the denominator 9 indicates that the original whole unit was divided into 9 pieces, and the numerator 23 indicates that we have 23 of these pieces. Dividing 9 into 23 will determine how many whole units we can obtain and how many ninths remain.

$$\frac{23}{9} = 2\frac{5}{9} \quad \textit{Steps:}$$
$$\text{9 into 23} = 2 \text{ remainder } 5$$

■ TECHNICAL TRADE APPLICATION: METEOROLOGY

The formula for converting a temperature in degrees Celsius (°C) to degrees Fahrenheit (°F) is:

$$°F = \left(\frac{9}{5} \times °C\right) + 32$$

If you were traveling in Europe and the temperature was 37 °C, what would that temperature be in degrees Fahrenheit?

$$°F = \left(\frac{9}{5} \times °C\right) + 32$$

$$= \left(\frac{9}{5} \times 37°C\right) + 32$$

$$= \left(\frac{9}{5} \times \frac{37}{1}\right) + 32$$

$$= \frac{333}{5} + 32$$

$$= 66\frac{3}{5} + 32$$

$$= 98\frac{3}{5}°F$$

SELF-TEST EVALUATION POINT FOR SECTION 1–5

Now that you have completed this section, you should be able to:

■ **Objective 11.** *Show how to multiply proper and mixed fractions.*

Use the following questions to test your understanding of Section 1–5.

1. $\frac{1}{3} \times \frac{2}{3} = ?$

2. $\frac{4}{8} \times \frac{2}{9} = ?$

3. $\frac{4}{17} \times \frac{3}{34} = ?$

4. $\frac{1}{16} \times 5\frac{3}{7} = ?$

5. $2\frac{1}{7} \times 7\frac{2}{3} = ?$

CHAPTER 1 / WHOLE NUMBERS AND FRACTIONS

Use the following technical trade questions to test your understanding of practical applications of Section 1–5.

1. **Masonry** Fourteen courses or rows of $2\frac{1}{2}$ inch bricks with $\frac{3}{8}$ inch mortar joints make up a residential perimeter wall (14 rows of bricks, 13 mortar joints). What is the height of the wall?
2. **Construction** If a red roof tile covers 7 inches ($\frac{7}{12}$ feet), how many feet of roof will be covered by 34 courses?
3. **Plumbing** The degree measure of a pipe bend is calculated by multiplying the fraction of the bend by 360 degrees. What is the measure of the following fraction bends, in degrees?

    a. $\frac{1}{4}$ c. $\frac{1}{5}$

    b. $\frac{1}{3}$

4. **Machining** A bolt has an 18-NF (National Fine) thread, which means that it will advance $\frac{1}{18}$ inch for every complete turn. How far will the bolt advance if a nut driver is used to turn it five complete revolutions?

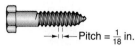
Pitch = $\frac{1}{18}$ in.

5. **Automotive** A hybrid (gasoline–electric) vehicle can average $32\frac{2}{5}$ miles per gallon. How many miles can it travel on $2\frac{1}{4}$ gallons?
6. **Carpentry** A $7\frac{3}{4}$ inch board is cut to two-thirds of its original thickness. What is its new size?

1–6   DIVIDING FRACTIONS

Once again let us examine what we are doing by using a simple example: How many quarters are in three-quarters? Stated mathematically, this problem appears as follows:

$$\frac{3}{4} \div \frac{1}{4} = ?$$

The answer is straightforward and requires no mathematical processes or steps because we already know that there are three quarters in three-quarters.

$$\frac{3}{4} \div \frac{1}{4} = 3$$

Since most problems are never this simple, we need to follow a mathematical process to divide fractions. Because addition is the opposite of subtraction, and multiplication is the opposite of division, we can just invert the second fraction in a division problem and then multiply both fractions to obtain the result. Let us try this process on the previous example:

$$\frac{3}{4} \div \frac{1}{4} = \frac{3 \times 4}{4 \times 1} = \frac{12 \div 4}{4 \div 4} = \frac{3}{1} = 3$$

Invert and multiply   Reduce   1 into 3 = 3

■ EXAMPLE

How many twelfths are in two-thirds $\left(\frac{2}{3} \div \frac{1}{12}\right)$?

■ *Solution:*

$$\frac{2}{3} \div \frac{1}{12} = \frac{2 \times 12}{3 \times 1} = \frac{24 \div 3}{3 \div 3} = \frac{8}{1} = 8$$

Invert and multiply   Reduce

1-6-1 Dividing Mixed Fractions

As when we multiply mixed fractions, to divide mixed fractions we must first convert all mixed fractions to improper fractions. Once we have done this, we can divide as usual, which means: Invert the second fraction and multiply. To start, let us use a simple example and test the procedure to see how many halves are in $2\frac{1}{2}$. As you have already determined, the result should indicate that there are five halves in two and one-half.

Steps
Convert mixed fraction to improper fraction:

$2 \times 2 = 4, 4 + 1 = 5.$ ⟶ $2\frac{1}{2} \div \frac{1}{2} = ?$

$\frac{5}{2} \div \frac{1}{2} = ?$

Invert second fraction and multiply. ⟶ $\frac{5 \times 2}{2 \times 1} = \frac{10}{2}$

Reduce. ⟶ $\frac{10 \div 2}{2 \div 2} = \frac{5}{1} = 5$

■ EXAMPLE

$6\frac{2}{3} \div 1\frac{1}{3} = ?$

■ Solution:

$6\frac{2}{3} \div 1\frac{1}{3} = ?$ *Steps:*
Convert mixed fractions to improper fractions:
$6 \times 3 = 18; 18 + 2 = 20.$
$\frac{20}{3} \div \frac{4}{3} = ?$ $1 \times 3 = 3; 3 + 1 = 4.$

$\frac{20 \times 3}{3 \times 4} = \frac{60}{12}$ Invert second fraction and multiply.

$\frac{60 \div 12}{12 \div 12} = \frac{5}{1} = 5$ Reduce.

Now let's do an example that does not work out so neatly.

■ EXAMPLE

How many times will $3\frac{5}{9}$ go into $6\frac{1}{2}$?

■ Solution:

Note that the term *go into* implies division and is asking how many times the first number will go into the second number.

$6\frac{1}{2} \div 3\frac{5}{9} = ?$ *Steps:*

$\frac{13}{2} \div \frac{32}{9} = ?$ Convert mixed fractions to improper fractions.

$$\frac{13 \times 9}{2 \times 32} = \frac{117}{64} \qquad \text{Invert second fraction and multiply.}$$

$$\frac{117}{64} = 1\frac{53}{64} \qquad \text{Convert improper fraction to mixed fraction.} \\ 117 \div 64 = 1 \text{ remainder } 53.$$

■ **TECHNICAL TRADE APPLICATION: PERSONAL FINANCE**

If you are paying an annual percentage rate of $15\frac{1}{8}$ percent on a credit card, what is the monthly percentage rate?

■ *Solution:*

$$15\frac{1}{8} \div 12 = ?$$

$$\frac{120}{8} \div \frac{12}{1} = ?$$

$$\frac{120 \times 1}{8 \times 12} = \frac{120}{96} = 1\frac{24}{96} = 1\frac{1}{4}\%$$

SELF-TEST EVALUATION POINT FOR SECTION 1–6

Now that you have completed this section, you should be able to:

■ *Objective 12.* *Show how to divide proper and mixed fractions.*

Use the following questions to test your understanding of Section 1–6.

1. $\dfrac{2}{3} \div \dfrac{1}{12} = ?$

2. $3\dfrac{1}{4} \div \dfrac{3}{4} = ?$

3. $8\dfrac{16}{20} \div 4\dfrac{4}{20} = ?$

4. $4\dfrac{3}{8} \div 2\dfrac{5}{16} = ?$

Use the following technical trade questions to test your understanding of practical applications of Section 1–6.

1. **Carpentry** How many $2\frac{3}{4}$ inch boards can be cut from a 16-foot plank?

2. **Electrical** How many $34\frac{1}{3}$ watt motors can be run on a 400 watt circuit?

3. **Machining** If a threaded pipe connecting a pool pump to an impeller has a $\frac{1}{16}$ inch thread pitch (16-NF), how many revolutions are needed to advance the pipe $2\frac{1}{2}$ inches?

4. **Printing** How many $2\frac{1}{4}$ foot pages can be cut from a 25 foot roll?

5. **Construction** How many $8\frac{1}{3}$ foot supporting beams can be cut from eight $28\frac{1}{2}$ foot sections of lumber?

6. **Drafting** If a $\frac{1}{2}$ inch to 1 foot scale $\left(\frac{1}{2}'' = 1'\right)$ is used on a set of plans, what size should a $16\frac{1}{4}$ foot long room be drawn as?

1–7 CANCELING FRACTIONS

Canceling is an operation that can be performed on fractions only when a multiplication symbol exists between the two. Its advantage is a reduction in the size of the numbers in the problem and therefore in time needed to solve, because fewer steps are needed. Let us demonstrate the process with a simple example.

Canceling
The process of removing a common divisor from numerator and denominator.

$$\frac{5}{12} \times \frac{4}{20} = ?$$

$$\frac{5 \times 4}{12 \times 20} = \frac{\boxed{20} \div 20}{240 \div 20} = \frac{1}{12}$$

As you can see from this example, multiplying large numbers generates large results that generally need to be reduced. By canceling before we perform the multiplication, we start with smaller fractions and thus simplify the process.

$$\frac{5}{\boxed{12}} \times \frac{\boxed{4}}{20} = \frac{\boxed{5}}{3} \times \frac{1}{\boxed{20}} = \frac{1}{3} \times \frac{1}{4} = \frac{1}{12}$$

Four divides evenly into 12 and 4.
$12 \div 4 = 3$,
$4 \div 4 = 1$.

Five divides evenly into 5 and 20.
$5 \div 5 = 1$,
$20 \div 5 = 4$.

Multiplication is now easier with no need for reducing.

Up to this point we have "reduced" fractions only by finding a number that will divide evenly into the top and bottom part of the same fraction. In this example you can see that we could not reduce $\frac{5}{12}$ because there was no number that would evenly divide into 5 and 12; however, we could reduce the numerator of one fraction with the denominator of another fraction. Reducing in this crisscross manner is called cancellation and can be done only when a multiplication sign is present between the two fractions.

■ EXAMPLE

Calculate the following: $\frac{4}{9} \div \frac{8}{27}$

■ Solution:

$$\frac{4}{9} \div \frac{8}{27} = ?$$

Invert second fraction and multiply.

Now that a multiplication symbol exists between the two fractions, we can see if it is possible to crisscross cancel.

$$\frac{\boxed{4}}{9} \times \frac{27}{\boxed{8}} = ?$$

Cancellation: $4 \div 4 = 1, 8 \div 4 = 2$.

$$\frac{1}{\boxed{9}} \times \frac{\boxed{27}}{2} =$$

Cancellation: $9 \div 9 = 1, 27 \div 9 = 3$.

$$\frac{1}{1} \times \frac{3}{2} = \frac{3}{2} = 1\frac{1}{2}$$

Convert improper fraction to mixed fraction.

■ EXAMPLE

Calculate the following: $3\frac{5}{7} \times 2\frac{2}{13}$

■ Solution:

$$3\frac{5}{7} \times 2\frac{2}{13} = ?$$

$$\frac{26}{7} \times \frac{28}{13} = \frac{2}{7} \times \frac{28}{1} = \frac{2}{1} \times \frac{4}{1}$$

Steps:
Convert mixed fractions to improper fractions:
$3 \times 7 = 21, 21 + 5 = 26$.
$2 \times 13 = 26, 26 + 2 = 28$

$$\frac{2}{1} \times \frac{4}{1} = \frac{8}{1} = 8$$

Cancellation: $13 \div 13 = 1, 26 \div 13 = 2$.
Cancellation: $28 \div 7 = 4, 7 \div 7 = 1$.

SELF-TEST EVALUATION POINT FOR SECTION 1–7

Now that you have completed this section, you should be able to:

■ **Objective 13.** *Demonstrate how to use cancellation to simplify fraction multiplication or division.*

Use the following questions to test your understanding of Section 1–7.

1. $\frac{16}{27} \times \frac{27}{240} = ?$
2. $\frac{8}{17} \times \frac{3}{4} = ?$
3. $\frac{15}{32} \div \frac{30}{4} = ?$
4. $6\frac{2}{3} \times 2\frac{7}{10} = ?$
5. $6\frac{1}{2} \div \frac{9}{32} = ?$

Use the following technical trade questions to test your understanding of practical applications of Section 1–6.

1. **Construction** As discussed previously, a roof's pitch can be expressed as a fraction:

$$\text{Pitch} = \frac{\text{Rise}}{\text{Run}} = \frac{12 \text{ feet}}{36 \text{ feet}}$$

Roofers will normally describe pitch as a single figure that is equal to the amount of rise for every 12 feet (denominator must be 12). Reduce the preceding pitch fraction to get an equivalent fraction with a denominator of 12.

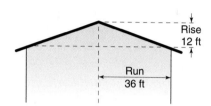

2. **Electrical** A 24 gauge copper wire measures $12\frac{6}{8}$ feet long. What is its length in lowest terms?

3. **Welding** When two 2 inch I-beams that measure $16\frac{3}{8}$ feet and $4\frac{1}{8}$ feet long are welded end-to-end, what is their combined length in lowest terms?

4. **Electrical** A web of five fiber-optic cable runs connects cameras to a control center. Add the following lengths to determine the total, and give it in lowest terms.

Cable length in feet: $8\frac{1}{2}, 40\frac{3}{8}, 24\frac{1}{4}, 18\frac{6}{16}, 16\frac{3}{4}$

5. **Automotive** A piston measures $3\frac{3}{32}$ inch when cold and $4\frac{27}{32}$ when hot. What is the difference in piston size in lowest terms?

6. **Metal Work** For a vent cover, a machinist cuts 4 strips of metal, each measuring $9\frac{7}{8}$ inch long. What is the total length of sheet metal used, in lowest terms?

REVIEW QUESTIONS

Multiple Choice Questions

1. What is the base or radix of the decimal number system?
   a. 100
   b. 10
   c. 0 through 9
   d. 10, 100, 1000, 10,000

2. Because each column in the decimal number system increases by _____ times as you move left, two columns to the left will be _____ times greater.
   a. 100, 10
   b. 10, 1000
   c. 1000, 10
   d. 10, 100

3. A fraction is a piece; therefore if you have a fraction, you have something _____.
   a. Between 1 and 10
   b. Greater than 1
   c. Less than 1
   d. Between 1 and 100

4. The more cuts you make in the whole unit, the _____ the fraction.
   a. Larger
   b. Smaller

5. The number above the fraction bar is called the _____, while the number below the fraction bar is called the _____.
   a. Denominator, factor
   b. Numerator, denominator
   c. Denominator, numerator
   d. Numerator, factor

6. How many ninths are in a whole unit?
   a. 9
   b. Depends on the value of the numerator
   c. 16
   d. All the above

7. If you had $6\frac{3}{4}$ apples in one pocket and $2\frac{1}{2}$ apples in the other, what would you have?
   a. Big pockets
   b. $8\frac{1}{4}$ apples
   c. $9\frac{3}{4}$ apples
   d. $9\frac{1}{4}$ apples

8. What is the lowest common denominator of 3 and 15?
   a. 45
   b. 15
   c. 5
   d. 90

9. A(n) _____ fraction has a numerator that is smaller than the denominator.
   a. Mixed
   b. Improper
   c. Proper
   d. All the above

10. A(n) _____ fraction contains both a whole number and a fraction.
    a. Mixed
    b. Proper
    c. Improper
    d. All the above

11. A(n) _____ fraction has a numerator that is larger than the denominator.
    a. Mixed
    b. Proper
    c. Improper
    d. All the above

12. Reduce the following fraction to its lowest terms: $\frac{32}{144}$.
    a. $\frac{4}{18}$
    b. $\frac{2}{9}$
    c. $\frac{1}{6}$
    d. $\frac{8}{32}$

Communication Skill Questions

13. Describe the base, positional weight, and reset and carry action of the decimal number system. (1–1)
14. What is a fraction, and what do the two numbers above and below the fraction bar line indicate? (1–2)
15. Describe the best method for calculating the lowest-value common denominator. (1–3)
16. What is a prime factor? (1–3)
17. What is an improper fraction? (1–3)
18. What is a mixed fraction? (1–3)
19. Arbitrarily choose values, and describe the steps used to:
    a. Add fractions. (1–3)
    b. Subtract fractions. (1–4)
    c. Multiply fractions. (1–5)
    d. Divide fractions. (1–6)
20. How can fractions be reduced? (1–3)
21. Why do you sometimes need to borrow when subtracting fractions? (1–4)
22. Arbitrarily choose values, and describe the steps used to:
    a. Add mixed fractions. (1–3)
    b. Subtract mixed fractions. (1–4)
    c. Multiply mixed fractions. (1–5)
    d. Divide mixed fractions. (1–6)
23. How are fractions canceled? (1–7)

Practice Problems

24. Indicate the positional weight of each of the following digits: 96,237.
25. Write each of the following values as a number.
    a. Two hundred and seventy-six
    b. Eight thousand and seven tens
    c. Twenty thousand, four tens, and nine units
26. Which columns will reset and carry when each of the following values is advanced or incremented by 1?
    a. 199
    b. 409
    c. 29,909
    d. 49,999
27. Write the whole number and fraction for each of the following quantities.

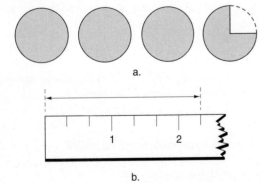

28. Consider the value $5\frac{7}{9}$. State which of the numbers represents the:
    a. Whole number
    b. Denominator
    c. Numerator
    d. Number of original pieces
    e. Number of fractional pieces we have

29. Add the following fractions.
    a. $\frac{5}{8} + \frac{2}{8} = ?$
    b. $\frac{47}{76} + \frac{15}{76} + \frac{1}{76} = ?$

30. Calculate the lowest common denominator and then add the following fractions.
    a. $\frac{4}{9} + \frac{1}{3} = ?$
    b. $\frac{4}{7} + \frac{2}{21} = ?$
    c. $\frac{2}{14} + \frac{1}{8} + \frac{1}{4} = ?$
    d. $\frac{4}{10} + \frac{1}{5} + \frac{2}{15} = ?$

31. Convert the following improper fractions to mixed fractions.
    a. $\frac{5}{2} = ?$
    b. $\frac{17}{4} = ?$
    c. $\frac{25}{16} = ?$
    d. $\frac{37}{3} = ?$

32. Add the following mixed fractions.
   a. $1\frac{3}{4} + 2\frac{1}{2} = ?$
   b. $9\frac{4}{10} + 6\frac{7}{8} = ?$

33. Reduce the following fractions.
   a. $\frac{4}{16} = ?$
   b. $\frac{16}{18} = ?$
   c. $\frac{74}{148} = ?$
   d. $\frac{28}{45} = ?$

34. Perform the following subtraction of fractions.
   a. $\frac{8}{14} - \frac{2}{7} = ?$
   b. $3\frac{4}{9} - 1\frac{1}{3} = ?$
   c. $4\frac{1}{16} - 2\frac{31}{32} = ?$
   d. $5\frac{3}{5} - 2\frac{4}{15} = ?$

35. Multiply the following. Cancel and reduce if possible.
   a. $\frac{1}{9} \times 4 = ?$
   b. $\frac{3}{6} \times \frac{4}{5} = ?$
   c. $\frac{1}{3} \times 4\frac{1}{2} = ?$
   d. $2\frac{3}{4} \times 4\frac{4}{11} = ?$

36. Divide the following. Cancel and reduce if possible.
   a. $\frac{1}{4} \div 8 = ?$
   b. $\frac{3}{5} \div \frac{5}{9} = ?$
   c. $3\frac{5}{7} \div 1\frac{6}{7} = ?$
   d. $14\frac{8}{9} \div 2\frac{4}{5} = ?$

Math Application Practice Problems

37. **Automotive** Referring to the gasoline gauge shown:
   a. How full is the tank?
   b. How much gas has been used?

38. **Construction** How deep is the sidewalk sinkhole shown?

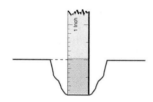

39. **Administration** Referring to the invoice work order shown, how much of the job (written as a fraction) has been completed.

	TIME NEEDED TO COMPLETE	TIME WORKED	AMOUNT COMPLETED
A	12 hours	5	?
B	7 hours	3	?

40. **Manufacturing** The control dial shown has to be turned from position y to position x by a production line operator. What is the distance between these two points as a fraction, both clockwise and counter-clockwise?

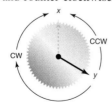

41. **Electronic** What is the wattage difference between the 1/8 watt resistor and 3/4 watt resistor shown?

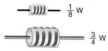

42. **Geography** The Earth has a surface area of approximately 196,800,000 square miles, of which, 3/4 is covered by water. Determine the number of square miles not covered by water.

43. **Automotive** List the order of the wrenches shown from smallest to largest.

44. **Waste Management** If the waste material shown in the two barrels was combined, what would be the total?

45. **Construction** A cement mixer, with a capacity of $9\frac{1}{2}$ cubic yards, makes 6 trips to the job site. How much concrete is delivered?

46. **Landscaping** A retaining wall is secured as shown below. How long should the bolts be?

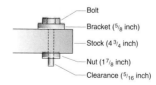

Web Site Questions

Go to the Web site http://www.prenhall.com/cook, select the textbook *Mathematics for the Technical Trades,* select this chapter, and then follow the instructions when answering the multiple-choice practice problems.

Decimal Calculation

An Apple a Day

One of the early pioneers who laid the foundations of many branches of science was Isaac Newton. He was born in a small farmhouse near Woolsthorpe in Lincolnshire, England, on Christmas Day in 1642. He was an extremely small, premature baby, which worried the midwives who went off to get medicine and didn't expect to find him alive when they came back. Luckily for science, however, he did survive.

Newton's father was an illiterate farmer who died three months before he was born. His mother married the local vicar soon after Newton's birth and left him in the care of his grandmother. This parental absence while he was growing up had a traumatic effect on him and throughout his life affected his relationships with people.

At school in the nearby town of Grantham, Newton showed no interest in classical studies but rather in making working models and studying the world around him. When he was in his early teens, his stepfather died and Newton had to return to the farm to help his mother. Newton proved to be a hopeless farmer; in fact, on one occasion when he was tending sheep he became so engrossed with a stream that he followed it for miles and was missing for hours. Luckily, a schoolteacher recognized Newton's single-minded powers of concentration and convinced his mother to let him return to school, where he performed better and later went off to Cambridge University.

In 1665, in his graduation year, Newton left Cambridge to return home to escape an epidemic of bubonic plague that had spread throughout London. During this time, Newton reflected on his years of seclusion at his mother's cottage and called them the most significant time in his life. It was here on a warm summer's day that Newton saw the apple fall to the ground, leading him to develop his laws of motion and gravitation. It was here that he wondered about the nature of light and later built a prism and proved that white light contains all the colors in a rainbow.

Later, when Newton returned to Cambridge, he demonstrated many of his discoveries but was reluctant to publish the details and did so finally only at the insistence of others. Newton went on to build the first working reflecting astronomical telescope and wrote a paper on optics that was fiercely challenged by the physicist Robert Hooke. Hooke quarreled bitterly with Newton over the years, and there were also heated debates about whether Newton or the German mathematician Gottfried Leibniz invented calculus.

The truth is that many of Newton's discoveries roamed around with him in the English countryside, and even though many of these would, before long, have been put forward by others, it was Newton's genius and skill (and long walks) that tied together all the loose ends.

Outline and Objectives

VIGNETTE: AN APPLE A DAY

INTRODUCTION

2–1 THE DECIMAL POINT

Objective 1: Define the function of the decimal point.

2–1–1 Positional Weight

Objective 2: Describe the positional weight of decimal fractions.

2–1–2 No Decimal Point and Extra Zeros

2–1–3 Converting Written Fractions to Decimal Fractions

Objective 3: Demonstrate how to:
    a. Convert written fractions to decimal fractions.
    b. Calculate reciprocals.
    c. Convert decimal fractions to written fractions.

2–1–4 Reciprocals

2–1–5 Converting Decimal Fractions to Written Fractions

2–2 CALCULATING IN DECIMAL

Objective 4: Describe the relationships among the four basic arithmetic operations: addition, subtraction, multiplication, and division.

2–2–1 Decimal Addition

Objective 5: Demonstrate how to perform the following arithmetic operations.
    a. Decimal addition
    b. Decimal subtraction
    c. Decimal multiplication
    d. Decimal division

2–2–2 Decimal Subtraction

2–2–3 Decimal Multiplication

2–2–4 Decimal Division

2–2–5 Ratios

Objective 6: Describe the following mathematical terms and their associated operations.
    a. Ratios
    b. Rounding off
    c. Significant places
    d. Percentages
    e. Averages

2–2–6 Rounding Off

2–2–7 Significant Places

2–2–8 Percentages

2–2–9 Averages

Introduction

The *decimal system of counting and calculating* was developed by Hindu mathematicians in India around A.D. 400. The system involved using fingers and thumbs, so it was natural that it would have 10 numerals or *digits*, which is a word meaning "finger." The use of this 10-digit or 10-finger system became widespread and by A.D. 800 was used extensively by Arab cultures. The 10 digits were represented by combinations of the 10 arabic numerals: *0, 1, 2, 3, 4, 5, 6, 7, 8,* and *9.* This is still called the *arabic number system.* The system eventually found its way to England, where it was adopted by nearly all the European countries by A.D. 1200. The Europeans called it the *decimal number system* after the Latin word *decimus,* which means "ten."

In this chapter we will discuss the decimal system of counting and calculating. **Decimal** is a topic with which you probably already feel extremely comfortable because we make use of it every day in countless applications. For example, our money is based on the decimal number system (there are 100 cents in 1 dollar, or 10 dimes in 1 dollar), and the metric system of measurement is also based on the decimal system (there are 10 millimeters in 1 centimeter, and 100 centimeters in 1 meter). Basic as it seems, however, you should still review this chapter because it will discuss all the terminology and procedures regarding how the decimal system is used, and this will serve as our foundation for all subsequent chapters.

Decimal
Numbered or proceeding by tens; based on the number 10.

2–1 THE DECIMAL POINT

Decimal Point
A period or centered dot between the parts of a decimal mixed number, separating the whole number from the fraction.

Decimal Fraction
A decimal value that is less than one.

The **decimal point (dp),** *symbolized by* "**.**", *is used to separate a whole decimal number from a* **decimal fraction.** Using a monetary example, we ask: What does the amount $12.75 actually mean? The number to the left of the decimal point, which is 12 in this example, indicates how many whole or complete dollars we have. On the other hand, the number to the right of the decimal point indicates the fraction, or the amount that is less than one dollar. Because there are 100 cents in $1, and this example indicates that we have 75 cents, this fraction is actually three-fourths of a dollar. In summary:

Decimal point

The numbers to the left of the decimal point indicate the amount of whole or complete units we have. In this monetary example, we have 12 complete or whole dollars.

$12.75

The numbers to the right of the decimal point indicate the *decimal fraction*, or amount that is less than one. In this monetary example, we have 75 cents or $\frac{75}{100}$ (which reduces to $\frac{3}{4}$).

The decimal point (dp) separates the whole numbers on the left from the fraction on the right. In this monetary example, the decimal point separates the complete dollars on the left from the fraction of a dollar or cents on the right.

Most calculators have a key specifically for entering the decimal point. It operates as follows.

CALCULATOR KEYS

Name: Decimal-point key

Function: Used in conjunction with the digit keys 0 through 9 to enter a decimal point. The position of the decimal point separates the whole number on the left from the decimal fraction on the right.

Example: Enter the number 12.75.

Press keys: [1] [2] [.] [7] [5]

Display shows: 12.75

2–1–1 *Positional Weight*

Most of us know that $0.50 is half a dollar, or 50 cents; $0.25 is a fourth or a quarter of a dollar, or 25 cents; and $0.75 is three-fourths or three-quarters of a dollar, or 75 cents; however, when it comes to explaining why $\frac{1}{2}$ = $0.50, $\frac{1}{4}$ = $0.25, and $\frac{3}{4}$ = $0.75, many people have difficulty. Why are these decimal fractions of 0.50, 0.25, and 0.75 equivalent to $\frac{1}{2}, \frac{1}{4}$, and $\frac{3}{4}$? The answer is in the position that each digit occupies in relation to the decimal point. To explain this further, Figure 2–1 repeats our previous example of $12.75, or twelve dollars and seventy-five cents. Once again, each place or position carries its own weight or value, and these positions are all relative to the decimal point. For instance, the digit in the first column to the left of the decimal point indicates how many units or ones are in the value, whereas the digit in the second column to the left of the decimal point indicates how many tens are in the value. Digits to the left of the decimal point all indicate the number of whole numbers in the value. On the other hand, digits to the right of the decimal point indicate the fraction or piece of a whole that is in the value. A value such as 3.5, therefore, indicates that we have three whole units and five-tenths $\left(\frac{5}{10}\right)$. Remembering our proper fractions from Chapter 1, we know that $\frac{5}{10}$ means that a whole unit was divided into ten parts and we have five of those parts. Because 5 is half of 10, we have half of a whole unit ($\frac{5}{10}$ can be reduced to $\frac{1}{2}$).

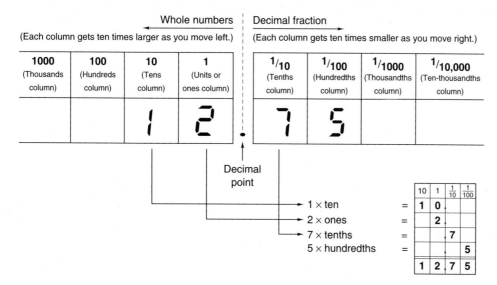

FIGURE 2–1 The Positional Weight of Decimal Fractions.

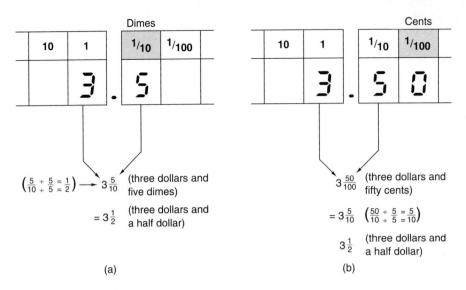

FIGURE 2–2 Interpreting the Decimal Fraction.

Is there a difference, therefore, between $3.5 and $3.50? The answer is no, which can be explained by looking at Figure 2–2. In Figure 2–2(a) we have interpreted the value as 3 dollars and $\frac{5}{10}$ of a dollar. Because $\frac{5}{10} = \frac{1}{2}$, the value in Figure 2–2(a) is 3 dollars and a half dollar. In Figure 2–2(b) we can interpret the decimal fraction as $\frac{5}{10}$ and $\frac{0}{100}$, or $\frac{50}{100}$. Because $\frac{50}{100} = \frac{5}{10} = \frac{1}{2}$, the value in Figure 2–2(b) is also 3 dollars and a half dollar. As far as money is concerned, we tend to refer to the fraction of a dollar in $\frac{1}{100}$ parts because a dollar is divided into 100 parts, and each of these parts is called a cent (there are 100 fractional parts or cents in a dollar).

Looking at this another way, we could label the $\frac{1}{10}$ column as dimes (10¢) and the $\frac{1}{100}$ column as cents (1¢). A value of 3.5 would then be interpreted as 3 dollars and 5 dimes, and a value of 3.50 would be interpreted as 3 dollars and 50 cents, as shown in Figure 2–2(a) and (b).

■ **EXAMPLE**

State the individual digit weights of the following number: 7.25.

■ *Solution:*

The first step is to give each column its respective weight or value.

1	1/10	1/100
7	2	5

It is now easier to see that we have 7 whole units, $\frac{2}{10}$ (two-tenths), and $\frac{5}{100}$ (five-hundredths). Said another way, we have 7 whole units and $\frac{25}{100}$ (twenty-five hundredths). Because $\frac{25}{100}$ can be reduced to $\frac{1}{4}$,

$$\frac{25 \div 25}{100 \div 25} = \frac{1}{4}$$

we could also interpret this value as seven and one-fourth $\left(7\frac{1}{4}\right)$.

2-1-2 No Decimal Point and Extra Zeros

What happens if a value such as

$$76$$

does not have a decimal point? The answer is: If you do not see a decimal point, always assume that it is to the right of the last digit in the number. In this example, therefore, the decimal point will follow the 6 in the number 76, as follows:

$$76.$$

Another point to remember: Always ignore extra zeros that are added to the left or right of a number. For example, to emphasize that a number such as .33 is a fraction, many people write this number as 0.33. The extra zero in this example is used to frame the decimal point so that it is not missed. You may have noticed that your calculator will add this zero if you press the keys $\boxed{.}\boxed{3}\boxed{3}$ and will display 0.33. Sometimes, zeros are added to the left of decimal whole numbers as well as to the right of decimal fractions. For example:

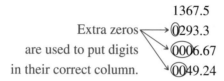

These extra zeros are used to align the digits of each number in the appropriate column. Without them, digits may be placed in the wrong column, as follows:

1367.5

0293.3

6.67

49.24

In other instances, extra zeros are placed to the right of a decimal fraction to indicate how to interpret the fraction. For example, three dollars and fifty cents would be written as 3.50 instead of 3.5 even though the extra zero after the five has no bearing on the value. This is so the value will be interpreted as three dollars and fifty cents $\left(\$3 + \frac{50}{100}\right)$ instead of three dollars and five dimes $\left(\$3 + \frac{5}{10}\right)$.

In all instances, extra zeros that are outside the number (to the left of the decimal whole number or to the right of the decimal fraction) do not change the value of the number.

2-1-3 Converting Written Fractions to Decimal Fractions

To convert proper fractions (such as $\frac{1}{2}, \frac{2}{6}, \frac{2}{3}, \frac{9}{16}$) to decimal fractions (such as 0.5, 0.3, 0.666, 0.562) we simply perform the operation that is indicated by the fraction bar. For example, when a fraction bar separates a numerator of 1 from a denominator of 2 $\left(\frac{1}{2}\right)$, the arithmetic operation described by the fraction bar is a division of 2 into 1. Performing this operation, you will convert the proper fraction $\frac{1}{2}$ to its decimal fraction equivalent.

$$\frac{1}{2} = 1 \div 2 = 0.5 \qquad \textit{Calculator sequence:}$$
$$\boxed{1}\boxed{\div}\boxed{2}\boxed{=}$$
$$\text{Answer: 0.5}$$

■ EXAMPLE

Convert the following proper fractions to decimal fractions.

a. $\frac{3}{4}$ b. $\frac{1}{4}$

c. $\dfrac{2}{3}$ e. $\dfrac{3}{9}$

d. $\dfrac{8}{16}$ f. $\dfrac{1}{8}$

■ *Solution:*

a. $\dfrac{3}{4} = 3 \div 4 = 0.75$

b. $\dfrac{1}{4} = 1 \div 4 = 0.25$

c. $\dfrac{2}{3} = 2 \div 3 = 0.666$

d. $\dfrac{8}{16} = 8 \div 16 = 0.5$ $\left(\dfrac{8 \div 8}{16 \div 8} = \dfrac{1}{2} = 0.5 \right)$

e. $\dfrac{3}{9} = 3 \div 9 = 0.333$ $\left(\dfrac{3 \div 3}{9 \div 3} = \dfrac{1}{3} = 0.3 \right)$

f. $\dfrac{1}{8} = 1 \div 8 = 0.125$

The next question is: How do we convert a mixed fraction such as $3\tfrac{1}{2}$ to its decimal equivalent? The answer is to do nothing with the whole number except place it in the whole-number decimal columns and then convert the proper fraction as before. Let us do the following example.

■ **EXAMPLE**

Convert the following mixed fractions to their decimal equivalents.

a. $3\dfrac{1}{2}$ c. $27\dfrac{4}{9}$

b. $486\dfrac{1}{4}$ d. $31{,}348\dfrac{8}{9}$

■ *Solution:*

a. $3\dfrac{1}{2} = 3.5$ $(1 \div 2 = 0.5)$

b. $486\dfrac{1}{4} = 486.25$ $(1 \div 4 = 0.25)$

c. $27\dfrac{4}{9} = 27.444$ $(4 \div 9 = 0.444)$

d. $31{,}348\dfrac{8}{9} = 31{,}348.888$ $(8 \div 9 = 0.888)$

Finally, let us convert an improper fraction such as $\tfrac{4}{2}$ (four halves) to its decimal equivalent. This is the easiest operation to perform, as shown in the following example, because all you do is perform the division on your calculator to obtain a decimal whole number and fraction.

■ **EXAMPLE**

Convert the following improper fractions to their decimal equivalents.

a. $\dfrac{4}{2}$

b. $\dfrac{7}{3}$

c. $6\dfrac{7}{2}$

d. $\dfrac{18}{16}$

■ *Solution:*

a. $\dfrac{4}{2} = 4 \div 2 = 2.0$

b. $\dfrac{7}{3} = 7 \div 3 = 2.333$

c. $6\dfrac{7}{2} = 6 + (7 \div 2) = 6 + 3.5 = 9.5 \quad \left(\dfrac{7}{2} = 3\dfrac{1}{2} = 3.5\right)$

d. $\dfrac{18}{16} = 18 \div 16 = 1.125 \quad \left(\dfrac{18 \div 2}{16 \div 2} = \dfrac{9}{8} = 1\dfrac{1}{8}\right)$

2-1-4 Reciprocals

Whenever you divide a number into 1, you get the **reciprocal** of that number. The word *reciprocal* means "inverse" or "opposite relation." This term can best be described by determining the reciprocal of two numbers, 2 and 2000. As just mentioned, to get the reciprocal of these numbers, we simply divide these numbers into 1, as shown:

Reciprocal
Inversely related.

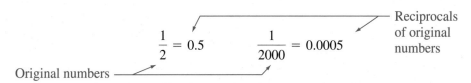

Comparing these two answers, you can see why the reciprocals, or result of dividing the numbers into 1, are inversely related to the original numbers. For instance, the reciprocal of a small number (↓), such as 2, results in a large fraction (↑), 0.5, whereas the reciprocal of a large number (↑), such as 2000, results in a small fraction (↓), 0.0005. This inverse relationship is even more evident when you consider that 2000 is one thousand times greater than 2, yet the reciprocal of 2000 is one thousand times smaller than the reciprocal of 2. To summarize:

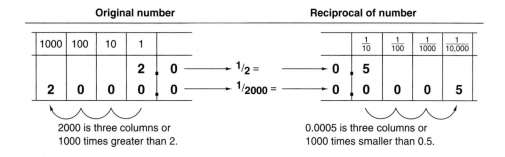

Most calculators have a key specifically for calculating the reciprocal of any number. It is called the reciprocal key and operates as follows.

CALCULATOR KEYS

Name: Reciprocal key

Function: Divides the value on the display into 1.

Example: $\dfrac{1}{3.2}$?

Press keys: [3] [.] [2] [1/x]

Display shows: 0.3125

Pi
The sixteenth letter of the Greek alphabet; the symbol π denoting the ratio of the circumference of a circle to its diameter; a transcendental number having a value to eight decimal places of 3.14159265.

The reciprocal of π is used in many different formulas. The Greek lowercase letter pi, which is pronounced "pie" and symbolized by π, is a constant value that describes how much bigger a circle's circumference is than its diameter.

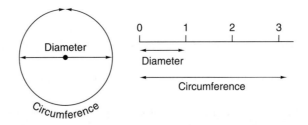

The approximate value of this constant is 3.142, which means that the circumference of any circle is approximately 3 times bigger than its diameter.

Most calculators have a key for quickly obtaining the value of pi. It is called the pi (π) key and operates as follows.

CALCULATOR KEYS

Name: Pi key

Function: Enters the value of pi correct to eight decimal digits.
$\pi = 3.1415927$

Example: $2 \times \pi = $?

Press keys: [2] [×] [π] [=]

Display shows: 6.2831853

CHAPTER 2 / DECIMAL CALCULATION

The factor π is used frequently in many formulas involved with the analysis of circular motion. The reciprocal $\frac{1}{2}\pi$ (1 over 2 × π) is frequently used, so we should do a few problems involving the reciprocal and pi keys on the calculator.

■ **EXAMPLE**

Calculate the reciprocals of the following numbers.

    a. 5   **b.** π **c.** 60   **d.** 2 × π

■ *Solution:*

    a. The reciprocal of $5 = \dfrac{1}{5} = 1 \div 5 = 0.2$.

         Calculator sequence: $\boxed{5}\,\boxed{1/x}$; display shows: 0.2

    b. The reciprocal of $\pi = \dfrac{1}{\pi} = \dfrac{1}{3.142} = 0.318$.

         Calculator sequence: $\boxed{\pi}\,\boxed{1/x}$; display shows: 0.3183098

    c. The reciprocal of $60 = \dfrac{1}{60} = 0.0166$.

         Calculator sequence: $\boxed{6}\,\boxed{0}\,\boxed{1/x}$; display shows: 0.0166667

    d. The reciprocal of $2 \times \pi = \dfrac{1}{2 \times \pi} = \dfrac{1}{6.28} = 0.159$.

         Calculator sequence: $\boxed{2}\,\boxed{\times}\,\boxed{\pi}\,\boxed{=}\,\boxed{1/x}$; display shows: 0.1591549.

2-1-5   *Converting Decimal Fractions to Written Fractions*

First, let us discuss why we would want to convert a decimal fraction, such as 0.5, into a written fraction, such as $\frac{1}{2}$. The answer is so that we can convert from a scale of 10 (decimal) to a scale that is something other than 10. For example, imagine that you have a standard ruler with 1 inch divided into 16 parts (there are sixteen $\frac{1}{16}$ in 1 inch). Next, imagine that you have to measure a length of 0.125 of an inch. Since the scale of the ruler is dealing with 16 parts and your decimal value of 0.125 is based on 10 parts, the two are not compatible and a conversion is needed. The first step is to convert $\frac{1}{16}$ into its decimal equivalent:

$$\frac{1}{16} = 1 \div 16 = 0.0625$$

This answer tells us that one-sixteenth of an inch is equivalent to 0.0625 of an inch in decimal. This can be proved by multiplying 0.0625 by 16 to see if we have 16 of these parts in 1 inch.

$$16 \times 0.0625 = 1 \text{ inch}$$

Now that we know that the value 0.0625 inch is equal to $\frac{1}{16}$ inch, the next step is to find out how many 0.0625 parts (or $\frac{1}{16}$) are in our decimal measurement of 0.125.

$$\frac{0.125}{0.0625} = 0.125 \div 0.0625 = 2$$

The answer 2 tells us that there are two 0.0625 parts (or sixteenths) in 0.125. Therefore, to measure off a length of 0.125 of an inch using a $\frac{1}{16}$ inch scale, simply measure a length of two-sixteenths $\left(\frac{2}{16}\right)$.

$$0.125 \text{ in.} = \frac{2}{16} \text{ in.}$$

■ **EXAMPLE**

Convert the following decimal fractions to proper fractions.

a. $0.25 = \dfrac{?}{8}$

b. $0.666 = \dfrac{?}{3}$

c. $0.59375 = \dfrac{?}{32}$

■ *Solution:*

a. 0.25 equals how many eighths? $\frac{1}{8}$ has a decimal equivalent of 0.125 ($1 \div 8 = 0.125$). How many 0.125 parts are in 0.25?

$$\dfrac{0.25}{0.125} = 0.25 \div 0.125 = 2$$

$$0.25 = \dfrac{2}{8} = \dfrac{1}{4}$$

b. $\dfrac{1}{3} = 0.333$

$$\dfrac{0.666}{0.333} = 2$$

$$0.666 = \dfrac{2}{3}$$

c. $\dfrac{1}{32} = 0.03125$

$$\dfrac{0.59375}{0.03125} = 19$$

$$0.59375 = \dfrac{19}{32}$$

CALCULATOR KEYS

Name: Conversion Functions
Display the answer as a fraction (▶ Frac).
Display the answer as a decimal (▶ Dec).

Function:
Value ▶ Frac (display as a fraction) displays an answer as its written fraction equivalent.
Value ▶ Dec (display as a decimal) displays an answer as its decimal equivalent.

Example: $\frac{1}{2} + \frac{1}{3}$ ▶ Frac

$$\dfrac{5}{6}$$

Ans ▶ Dec

.8333333333

CHAPTER 2 / DECIMAL CALCULATION

SELF-TEST EVALUATION POINT FOR SECTION 2–1

Now that you have completed this section, you should be able to:

- **Objective 3.** Define the function of the decimal point.
- **Objective 4.** Describe the positional weight of decimal fractions.
- **Objective 5.** Demonstrate how to convert written fractions to decimal fractions, calculate reciprocals, and convert decimal fractions to written fractions.

Use the following questions to test your understanding of Section 2–1.

1. What are the individual digit weights of the number 178.649?
2. Convert the following written fractions to decimal fractions.
   a. $1\frac{5}{2}$ b. $192\frac{3}{4}$ c. $67\frac{6}{9}$
3. Calculate the reciprocal of the following numbers.
   a. 2500   b. 0.25   c. 0.000025
4. Convert the following decimal fractions to written fractions.
   a. $7.25 = \frac{?}{16}$ b. $156.90625 = \frac{?}{32}$

Use the following technical trade questions to test your understanding of practical applications of Section 2–1.

1. **Health Care** A capsule contains 0.25 grams of potassium bromide. How many capsules could be made with one gram of potassium bromide?
2. **Machining** Convert the following drill bit diameters to decimal equivalent values:
   a. $\frac{3}{4}$
   b. $\frac{5}{16}$
   c. $\frac{3}{8}$
   d. $\frac{5}{32}$
3. **Drafting** A dimension on a set of plans measures $3\frac{1}{4}$ inches. What would this be as a decimal equivalent value?
4. **Electrical** Three sections of 18-gauge Romex cable measure $18\frac{1}{4}$ feet, $6\frac{1}{3}$ feet, and $24\frac{1}{8}$ feet. Convert these to their decimal equivalent values.
5. **Metal Work** Convert the thicknesses of the following pieces of sheet metal to their fractional equivalents in 64ths:
   a. 0.046875 inch
   b. 0.859375 inch
6. **Automotive** A main cylinder measures $2\frac{9}{16}$ inches in diameter before it is rebored to increase its size by 0.05 inch. What is the new size of the cylinder as a decimal value?

2–2 CALCULATING IN DECIMAL

In this section we first review the basic principles of decimal addition, subtraction, multiplication, and division. As you will discover, there are really only *two basic operations: addition and subtraction*. If you can perform an addition, you can perform a multiplication because *multiplication is simply repeated addition*. For example, a problem such as 4×2 (4 multiplied by 2) is asking you to calculate the sum of four 2s ($2 + 2 + 2 + 2$). Similarly, *division is simply repeated subtraction*. For example, a problem such as $9 \div 3$ (9 divided by 3) is asking you to calculate how many times 3 can be subtracted from 9 ($9 - 3 = 6$, $6 - 3 = 3$, $3 - 3 = 0$; therefore, there are three 3s in 9). In summary, therefore, the two basic mathematical operations (addition and subtraction) and their associated mathematical operations (multiplication and division) are as follows:

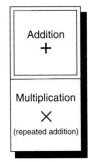
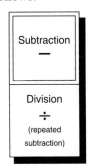

There is another relationship that needs to be discussed, and that is the inverse or opposite nature between addition and subtraction and between multiplication and division. For example, *subtraction is the opposite arithmetic operation of addition.* This can be proved by starting with a number such as 10, then adding 5, then subtracting 5.

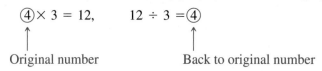

⑩ + 5 = 15,   15 − 5 = ⑩

    ↑   ↑

Original number   Back to original number

Similarly, *multiplication is the opposite arithmetic operation of division.* This can be proved by starting with a number such as 4, then multiplying by 3, then dividing by 3.

④ × 3 = 12,   12 ÷ 3 = ④

   ↑   ↑

Original number   Back to original number

In summary, therefore, addition is the opposite mathematical operation of subtraction, and multiplication is the opposite mathematical operation of division. This opposite relationship will be made use of in a subsequent chapter.

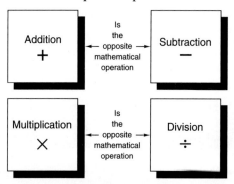

Every calculator has keys for performing these four basic arithmetic operations. The name, function, and procedure for using these keys are as follows.

CALCULATOR KEYS

Name: Decimal point key [.] (discussed previously)

Name: Equals key [=]

Function: Combines all previously entered numbers and operations. Used to obtain intermediate and final results.

Example: 3 × 4 + 6 = ?

Press keys: [3] [×] [4] [=] [+] [6] [=]

Display shows: 18.0

Name: Add key [+]

Function: Instructs calculator to add the next entered quantity to the displayed number.

Example: Add 31.6 + 2.9 = ?

Press keys: [3] [1] [.] [6] [+] [2] [.] [9] [=]

Displays shows: 34.5

Name: Subtract key [−]

Function: Instructs calculator to subtract the next entered quantity from the displayed number.

> **CALCULATOR KEYS** (continued)
>
> **Example:** 266.3 − 5.35 = ?
> Press keys: 2 6 6 . 3 − 5 . 3 5 =
> Display shows: 260.95
>
> **Name:** Multiply key ×
>
> **Function:** Instructs calculator to multiply the displayed number by the next entered quantity.
>
> **Example:** 3.85 × 2.9 = ?
> Press keys: 3 . 8 5 × 2 . 9 =
> Display shows: 11.165
>
> **Name:** Divide key ÷
>
> **Function:** Instructs calculator to divide the displayed number by the next entered quantity.
>
> **Example:** 28.98 ÷ 2.3 = ?
> Press keys: 2 8 . 9 8 ÷ 2 . 3 =
> Display shows: 12.6

2–2–1 Decimal Addition

Addition is the process of gathering together two or more numbers into one total sum.

■ **EXAMPLE**

Add 3 and 2.

■ *Solution:*

$$3 + 2 = 5$$

You have probably noticed that the order which numbers are added has no effect on the final result or sum. To use the values in the previous example, whether you add 3 + 2 or 2 + 3, the result will always be 5. This property of addition is formally known as the *commutative law of addition*.

$$a + b = b + a$$

Commutative Law of Addition: The order in which you add two numbers will have no effect on the sum.

When three numbers are added together, we generally add two numbers first to obtain a partial sum, and then add this to the remaining number.

■ **EXAMPLE**

Add 3 and 7 and 4.

■ *Solution:*

Steps:

$$3 + 7 + 4 = \quad\quad 3 + 7 = 10$$
$$10 + 4 = \quad\quad 10 + 4 = 14$$
$$14$$

From the previous example, notice that whichever two numbers you group and add first has no effect on the final sum. For example,

Parentheses are used to indicate which group is to be added first.

$(3 + 7) + 4 =$
$10 + 4 =$
14

$3 + (7 + 4) =$
$3 + 11 =$
14

This property is known as the *associative law of addition.*

$$(a + b) + c = a + (b + c)$$

Addition
The operation of combining numbers so as to obtain an equivalent quantity.

Associative Law of Addition: The grouping of the numbers you add first will have no effect on the final sum.

To review decimal **addition,** let us examine the steps involved in adding the following numbers.

$$66.332 + 285 + 0.002 + 182.788 = ?$$

First, list the numbers vertically, making sure that the decimal points are lined up as follows.

Addition process:

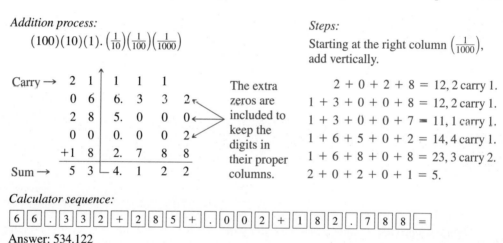

Steps:

Starting at the right column $\left(\frac{1}{1000}\right)$, add vertically.

$2 + 0 + 2 + 8 = 12, 2$ carry 1.
$1 + 3 + 0 + 0 + 8 = 12, 2$ carry 1.
$1 + 3 + 0 + 0 + 7 = 11, 1$ carry 1.
$1 + 6 + 5 + 0 + 2 = 14, 4$ carry 1.
$1 + 6 + 8 + 0 + 8 = 23, 3$ carry 2.
$2 + 0 + 2 + 0 + 1 = 5$.

Calculator sequence:

| 6 | 6 | . | 3 | 3 | 2 | + | 2 | 8 | 5 | + | . | 0 | 0 | 2 | + | 1 | 8 | 2 | . | 7 | 8 | 8 | = |

Answer: 534.122

Carrying
The process of transferring from one column to another during an addition process.

Sum
The result obtained when adding numbers.

There are a few points to note regarding this previous addition. First, the number 285 did not have a decimal point indicated; therefore, it was assumed to be to the right of the last digit (285.0). Second, what is really happening when we **carry** or overflow into the next-left column? The answer is best explained by examining the addition of the units column in the previous example. In this column we had a result of $1 + 6 + 5 + 0 + 2 = 14$. As you know from our discussion on positional weight, 14 can be broken down into 1 ten (1×10) and 4 ones (4×1). This is why the 1 ten is carried over to the tens column, and the 4 ones or units are placed in the units sum column. Explained another way, whenever the sum of all the digits in any column exceeds the maximum single decimal digit 9, a reset and carry action occurs. In the case of the units column with a sum of 14, we exceeded 9 and therefore the units sum was reset to 0 and then advanced to 4, and a 1 was carried into the tens column to make 14.

In summary, therefore, because each column to the left is 10 times larger, a reset and carry action will occur every time the sum of any column is incremented or advanced by one, from 9 to 10.

■ EXAMPLE

How much is $976.73 + $998.28?

■ Solution:

Carry:	1111 1
Augend:	0976.73
Addend:	+ 0998.28
Sum:	$1975.01

Steps:

$3 + 8 = 11$, 1 carry 1.
$1 + 7 + 2 = 10$, 0 carry 1.
$1 + 6 + 8 = 15$, 5 carry 1.
$1 + 7 + 9 = 17$, 7 carry 1.
$1 + 9 + 9 = 19$, 9 carry 1.
$1 + 0 + 0 = 1$, 1 carry 0.

Augend
The number to which a new quantity, called the addend, is to be added.

Addend
The number or quantity to be added to an existing quantity, called the augend.

Calculator sequence: [9][7][6][.][7][3][+]
[9][9][8][.][2][8][=]

Answer: 1975.01

■ TECHNICAL TRADE APPLICATION: ACCOUNTING

Calculate the total cost of the following service station repair invoice:

Parts:	$ 64.95
Labor:	$226.35
Coolant:	$ 4.25
Tax:	$ 22.17

■ Solution:

```
     64.95
    226.35
      4.25
   + 22.17
   $317.72
```

■ TECHNICAL TRADE APPLICATION: PERSONAL FINANCE

How many hours did you work in the previous week if your time card shows the following:

Monday—6 hours
Tuesday—4.5 hours
Wednesday—7.5 hours
Thursday—5 hours
Friday—8 hours

■ Solution:

$6 + 4.5 + 7.5 + 5 + 8 = 31$ hours

2–2–2 Decimal Subtraction

Subtraction is the process of finding the difference between two numbers by taking away the lesser number from the greater number.

■ **EXAMPLE**

Are all of the following subtractions?

- **a.** Subtract 4 from 7.
- **b.** What is the difference between 4 and 7?
- **c.** How much greater is 7 than 4?
- **d.** How much less is 4 than 7?

■ *Solution:*

Yes, these are all different ways of describing subtraction, and in all cases:

$$7 - 4 = 3$$

To review decimal **subtraction**, let us examine the steps involved in the following example:

$$\$2005.23 - \$192.41 = ?$$

To begin, list the numbers vertically, making sure that the decimal points are lined up as follows:

Subtraction process:

```
                                    Dimes  Cents
                                      ↓      ↓
                (1000)(100)(10)(1)  (1/10) (1/100)
Borrow:              1    9           4
Minuend:         2¹⁰0¹⁰5.¹2    3
Subtrahend:      0   1   9   2.  4    1
Difference:      1   8   1   2.  8    2
```

Calculator sequence:

| 2 | 0 | 0 | 5 | . | 2 | 3 | − |
| 1 | 9 | 2 | . | 4 | 1 | = |

Answer: 1812.82
   or $1812.82

Steps:
Starting at the right column ($\frac{1}{100}$ or cent) subtract vertically:
$3 - 1 = 2$
$2 - 4$: impossible to subtract 4 from the smaller number 2. Therefore, we will *borrow* $1 from the $5 in the units column. This will leave $4 in the units column; because $1 = 10 dimes, we will be adding 10 dimes to the 2 dimes already present, giving a total of 12 dimes.
$12 - 4 = 8$
$4 - 2 = 2$
$0 - 9$: impossible to subtract 9 from nothing or zero.
Because we cannot borrow from the tens or hundreds columns, we will have to borrow from the thousands column.
Borrowing $1000(10 \times 100)$ gives 10 hundreds; borrowing 1 hundred from 10 hundreds gives 9 hundreds. Borrowing $100(10 \times 10)$ gives 10 tens.
$10 - 9 = 1$
$9 - 1 = 8$
$1 - 0 = 1$

Subtraction
The operation of deducting one number from another.

Minuend
A number from which the subtrahend is to be subtracted.

Subtrahend
A number that is to be subtracted from a minuend.

Difference
The amount by which things differ in quantity or measure.

As you can see from this example, *borrowing* is the exact opposite of carrying. With addition we sometimes end up with a sum of 10 or more, which results in an overflow or carry into the next-higher or next-left column, whereas with subtraction we sometimes end up with a larger number in the subtrahend than the minuend, and it becomes necessary to borrow a 1 from the next-higher place to the left in the minuend.

■ EXAMPLE

Perform the following subtraction:

$$27{,}060.0143 - 963.86$$

■ Solution:

	(10,000)	(1000)	(100)	(10)	(1)	($\frac{1}{10}$)	($\frac{1}{100}$)	($\frac{1}{1000}$)	($\frac{1}{10{,}000}$)	
		6	9	¹5	9.	9				*Steps:*
	2	7̸	1̸0̸	6	1̸0̸.	1̸0̸	¹1	4	3	$3 - 0 = 3$
	−0	0	9	6	3.	8	6	0	0	$4 - 0 = 4$
Difference:	2	6	0	9	6.	1	5	4	3	$1 - 6$: can't do; borrow.

Calculator sequence:

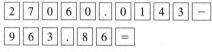

Answer: 26,096.1543

Borrowing 1 ten from 6 tens gives 5 tens and 10 ones. Borrowing 1 one from 10 ones gives 9 ones and 10 tenths.
Borrowing 1 tenth from 10 tenths gives 9 tenths and 10 hundredths.
10 hundredths + the original 1 hundredth = 11 hundredths:
$11 - 6 = 5$
$9 - 8 = 1$
$9 - 3 = 6$
$5 - 6$: can't do; borrow.
Borrowing 1 thousand from 7 thousands gives 6 thousands and 10 hundreds.
Borrowing 1 hundred from 10 hundreds gives 9 hundreds and 10 tens. 10 tens + the original 5 tens = 15 tens:
$15 - 6 = 9$
$9 - 9 = 0$
$6 - 0 = 6$
$2 - 0 = 2$

■ TECHNICAL TRADE APPLICATION: RETAIL

If the regular price of a television is $856.99 and the sale price is $699.99, how much will you save?

■ Solution:

$$\begin{array}{r} \$859.99 \\ -\$699.99 \\ \hline \$160.00 \end{array}$$

A savings of $160.00.

SECTION 2–2 / CALCULATING IN DECIMAL

■ **TECHNICAL TRADE APPLICATION: OFFICE MANAGEMENT**

If an employee's gross weekly salary is $695.00 and her deductions total $212.65, what is her take-home pay?

■ *Solution:*

$$\begin{array}{r} \$695.00 \\ -\$212.65 \\ \hline \$482.35 \end{array}$$ ← Her weekly take-home pay is $482.35.

2–2–3   *Decimal Multiplication*

Multiplication is a process in which a number is added to itself a given number of times. The given number is called the *multiplicand*, the number that specifies how many times the number is to be added to itself is called the *multiplier*, and the result is called the *product*.

■ **EXAMPLE**

What is the product of 3 times 2?

■ *Solution:*

$$3 \times 2 = 6$$

Two threes $(3 + 3)$ are six.

$$\begin{array}{r} 3 \\ \times 2 \\ \hline 6 \end{array} \begin{array}{l} \text{Multiplicand} \\ \text{Multiplier} \\ \text{Product} \end{array}$$

The commutative law also applies to multiplication, in that the order in which you multiply two numbers will have no effect on the result. To use the values in the previous example, whether you multiply 2×3 or 3×2, the result will in both cases be 6. This property of multiplication is known as the *commutative law of multiplication*.

$$\boxed{a \times b = b \times a}$$

Commutative Law of Multiplication: The order in which you multiply two numbers will have no effect on the product.

When three numbers are to be multiplied, we generally multiply one number by another to obtain a partial product, and then multiply this result by the final number.

■ **EXAMPLE**

What is the result of 2 times 3 times 10?

■ *Solution:*

$$2 \times 3 \times 10 = ?$$

No matter which pair is multiplied first, the result is still the same.

$$\begin{array}{ll} 2 \times (3 \times 10) = & (2 \times 3) \times 10 = \\ 2 \times 30 = & 6 \times 10 = \\ 60 & 60 \end{array}$$

CHAPTER 2 / DECIMAL CALCULATION

The previous example confirms the *associative law of multiplication.*

$$a \times (b \times c) = (a \times b) \times c$$

Associative Law of Multiplication: The grouping of the numbers you multiply first will have no effect on the final product.

To review decimal **multiplication,** let us examine the steps involved in the following example:

$$27.66 \times 2.4 = ?$$

Unlike in the previous mathematical operations, we can ignore the alignment of the decimal point and simply make sure that the two numbers are flush on the right side.

Multiplicand: 27.66
Multiplier: × 2.4

Two numbers are listed vertically so that they are flush on the right side.

We can now use the following multiplication steps:

Multiplication process:

Second carry:	11 1
First carry:	132 2
Multiplicand:	27.6 6
Multiplier:	× ②④
First partial product:	110 6 4
Second partial product:	553 2 0
Final product:	663 8 4

Steps:
To begin, multiply each digit in the multiplicand by the first digit of the multiplier (which is 4) to obtain the first partial product. Line up the rightmost digit of the first partial product under the multiplier digit being used (see arrow).

$4 \times 6 = 24$, 4 carry 2.
$4 \times 6 = 24$, $24 +$ previous carry of $2 = 26$, 6 carry 2.
$4 \times 7 = 28$, $28 +$ carry of $2 = 30$, 0 carry 3.
$4 \times 2 = 8$, $8 +$ carry of $3 = 11$, 1 carry 1; bring 1 down.

The next step is to multiply each digit in the multiplicand by the second digit of the multiplier (which is 2) to obtain the second partial product. Line up the rightmost digit of the second partial product under the multiplier digit being used (see arrow).

$2 \times 6 = 12$, 2 carry 1.
$2 \times 6 = 12$, $12 +$ carry of $1 = 13$, 3 carry 1.
$2 \times 7 = 14$, $14 +$ carry of $1 = 15$, 5 carry 1.
$2 \times 2 = 4$, $4 +$ carry of $1 = 5$.

The next step is to add the first and second partial products to obtain the final product.
$4 + 0 = 4$
$6 + 2 = 8$
$0 + 3 = 3$
$1 + 5 = 6$
$1 + 5 = 6$

Calculator sequence:

| 2 | 7 | . | 6 | 6 | × | 2 | . | 4 | = |

Answer: 66.384

Multiplication
The operation of adding an integer to itself a specified number of times.

Multiplicand
The number that is to be multiplied by another.

Multiplier
One that multiplies; a number by which another number is multiplied.

27.6̂6̂ The multiplicand has two digits to the right of the decimal point.
2.4 The multiplier has one digit to the right of the decimal point.
2 + 1 = 3 places.

← The final step is to determine the position of the decimal point in the answer by counting the number of digits to the right of the decimal point in the multiplier and multiplicand. There are three in this example, so the final product will have three digits to the right of the decimal point.
66.3̂ 8̂ 4̂

Answer is 66.384

■ EXAMPLE

How much is 0.035 multiplied by 0.005?

■ Solution:

In this example we will take a few shortcuts. Since the decimal points do not make a difference at this stage, let us drop them completely. This will produce:

$$\begin{array}{r} 0035 \\ \times\ 0005 \\ \hline \end{array}$$

Second, we learned earlier that extra zeros appearing to the left or right of a number are not needed in the multiplication process, so let us remove these zeros:

$$\begin{array}{r} 35 \\ \times\ 5 \\ \hline \end{array}$$

The multiplication is now a lot easier because it is simply 35 × 5. Performing the multiplication, we obtain the product:

$$\begin{array}{r} \overset{2}{3}5 \\ \times\ \ 5 \\ \hline 175 \end{array}$$

The final step is to determine the position of the decimal point. The original multiplicand had three digits to the right of the decimal point (0.0̂ 3̂ 5̂), and the original multiplier also had three digits to the right of the decimal point (0.0̂ 0̂ 5̂), so the final product should have six digits to the right (3 + 3) of the decimal point.

.0̂ 0̂ 0̂ 1̂ 7̂ 5̂   or   0.000175

■ TECHNICAL TRADE APPLICATION: AUTOMOTIVE

If your car averages 26.3 miles per gallon, how far can you travel if your gas tank is filled with 12 gallons of gasoline?

■ Solution:

26.3 miles per gallon × 12 gallons = 315.6 miles

■ TECHNICAL TRADE APPLICATION: SURVEYING

If an inch on a street map is equivalent to 0.5 mile, how many miles will you have to travel if the distance on the map measures 14.5 inches?

■ Solution:

14.5 inches × 0.5 mile per inch = 7.25 miles

2-2-4 *Decimal Division*

To review decimal **division** (or *long division*, as it is sometimes called) let us step through the following division example: 1362 ÷ 12. The procedure is as follows.

Long-division process:

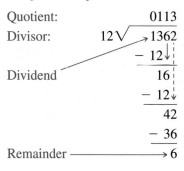

Steps:

The first step is to see how many times the divisor (12) can be subtracted from the first digit of the dividend (1). Because the dividend is smaller, the answer or quotient is zero (1 − 12 cannot be done). We place this zero above the dividend 1.

Next, we see how many times the divisor (12) can be subtracted from the first two digits of the dividend (13). 13 − 12 = 1, remainder 1. The answer or quotient (1) is placed above the next digit of the dividend (3), and the remainder of the subtraction (1) is dropped below the 13 − 12 line.

We carry the next digit of the dividend (6) down to the remainder (1) to address the next part of the dividend, which is 16. Now we repeat the process by seeing how many times the divisor, 12, can be subtracted from these two digits, 16. The answer is a quotient of 1 (which is placed above the 6 dividend) with a remainder of 4 (which is dropped below the 16 − 12 line).

Division
The operation of separating a number into equal parts.

Quotient
The number resulting from the division of one number by another.

Divisor
The number by which a dividend is divided.

Dividend
A number to be divided.

Remainder
The final undivided part after division that is less or of lower degree than the divisor.

Calculator sequence:

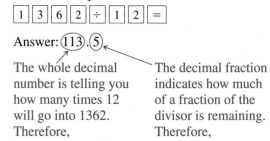

Answer: (113).(5)

The whole decimal number is telling you how many times 12 will go into 1362. Therefore,
1362 ÷ 12 = (113)

The decimal fraction indicates how much of a fraction of the divisor is remaining. Therefore,
0.5 × 12 = 6

Answer to 1362 ÷ 12 is 113, remainder 6.

As another example, 13 ÷ 4 = ?

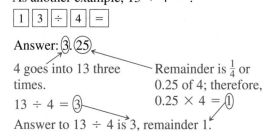

Answer: (3).(25)

4 goes into 13 three times.
13 ÷ 4 = (3)

Remainder is $\frac{1}{4}$ or 0.25 of 4; therefore,
0.25 × 4 = (1)

Answer to 13 ÷ 4 is 3, remainder 1.

Bringing down the next dividend digit, 2, and combining it with the previous remainder, 4, we now have the final part of the dividend, which is 42. Because 12 can be subtracted from 42 three times, a quotient of 3 is placed above the final digit of the dividend (2), and the remainder of the subtraction, 6, is dropped below the 42 − 36 line.

Therefore,
1362 ÷ 12 = 113, remainder 6.
Said another way, 12 can be subtracted from 1362 113 times, with 6 remaining.

■ **EXAMPLE**

Calculate the result of dividing 19,205.628 by 1.56.

■ *Solution:*

19,205.628 ÷ 1.56 = 12,311.3

Calculator sequence: [1][9][2][0][5][.][6][2][8][÷][1][.][5][6][=]

Answer: 12,311.3

■ **TECHNICAL TRADE APPLICATION: PERSONAL FINANCE**

If your round-trip commute to work for five days totals 87.5 miles, how many miles are you driving per day?

■ *Solution:*

87.5 miles ÷ 5 days = 17.5 miles per day

■ **TECHNICAL TRADE APPLICATION: PERSONAL FINANCE**

If your gross pay for a week is $536.50 and you worked 37 hours, what is your hourly rate of pay?

■ *Solution:*

$536.50 ÷ 37 hours = $14.50 per hour

2-2-5   *Ratios*

Ratio
The relationship in quantity, amount, or size between two or more things.

A **ratio** is a comparison of one number with another number. For example, a ratio such as 7 : 4 ("seven to four") is comparing the number 7 with the number 4. This ratio could be written in one of the following three ways:

$\frac{7}{4}$ ← Ratio is written as a fraction.

7 : 4 ← Ratio is written using the ratio sign (:).

1.75 to 1 ← Ratio is written as a decimal.

To express the ratio as a decimal, the number 7 was divided by the number 4, giving a result of 1.75 (7 ÷ 4 = 1.75). This result indicates that the number 7 is 1.75 times (or one and three-quarter times) larger than the number 4.

■ **TECHNICAL TRADE APPLICATION: RETAIL**

If one store has 360 cell phones and another store has 100, express the ratio of 360 to 100:

a. Using the divide sign
b. Using the ratio sign
c. As a decimal

■ *Solution:*

a. $\frac{360}{100} = \frac{360 \div 20}{100 \div 20} = \frac{18}{5}$ ← 360 to 100 and 18 to 5 are equivalent ratios.

b. 360 : 100 = 18 : 5
c. 3.6 ← The first store has 3.6 times as many items as the other store (18 ÷ 5 = 3.6).

■ **EXAMPLE**

Calculate the ratio of the following two like quantities. If necessary, reduce the ratio to its lowest terms.

    a. $0.8 \div 0.2 = ?$
    b. $0.0008 \div 0.0002 = ?$

■ *Solution:*

    a. How many 0.2s are in 0.8? The answer is the same as saying, How many 2s are in 8?

$$0.8 \div 0.2 = 4 \qquad \frac{8 \div 2}{2 \div 2} = \frac{4}{1} \quad \text{(a ratio of 4 to 1)}$$

    b. How many 0.0002s are in 0.0008?

$$0.0008 \div 0.0002 = 4:1$$

The ratio in example (a) is the same as in (b). Another way to describe this is to say that because the size of each decimal fraction compared with the other decimal fraction (or the *ratio*) remained the same, the answer or quotient remained the same.

    8 is four times larger than 2, or 2 is four times smaller than 8.

    0.8 is four times larger than 0.2, or 0.2 is four times smaller than 0.8.

    0.0008 is four times larger than 0.0002, or 0.0002 is four times smaller than 0.0008.
    Therefore, all have the same $4:1$ ratio.

■ **TECHNICAL TRADE APPLICATION: ADMINISTRATION**

In some instances, a ratio will be used to describe how a quantity is to be divided. For example, if 600 textbooks are to be divided between two schools in the ratio of 1 to 2 ($1:2$), how many textbooks will each school get? To calculate this, you should first add the terms of the ratio:

■ *Solution:*

    Step 1: The ratio is $1:2$, so $1 + 2 = 3$ (this means that the total is to be divided into 3 parts).

Next, you should multiply the total by each fractional part:

    Step 2: The first school will receive $\frac{1}{3}$ of the total 600 textbooks, which is 200 texts $\left(\frac{1}{3} \times 600 = 200\right)$. The second school will receive $\frac{2}{3}$ of the total 600 textbooks, which is 400 texts $\left(\frac{2}{3} \times 600 = 400\right)$.

■ **TECHNICAL TRADE APPLICATION: FOOD SERVICE**

If three restaurants are shipped 216 hamburgers in the ratio of $6:14:7$, how many hamburgers will each restaurant get?

■ *Solution:*

The first step is to add the terms of the ratio:

    Step 1: The ratio is $6:14:7$, so $6 + 14 + 7 = 27$ (this means that the total is to be divided into 27 parts).

Next, you should multiply the total by each fractional part:

    Step 2: The first restaurant will receive $\frac{6}{27}$ of the total 216 hamburgers, which is 48 $\left(\frac{6}{27} \times 216 = 48\right)$. The second restaurant will receive $\frac{14}{27}$ of the total 216 hamburgers, which is 112 $\left(\frac{14}{27} \times 216 = 112\right)$. The third restaurant will receive $\frac{7}{27}$ of the total 216 hamburgers, which is 56 $\left(\frac{7}{27} \times 216 = 56\right)$.

Rounding Off

An operation in which a value is abbreviated by applying the following rule: When the first digit to be dropped is 6 or more, or 5 followed by a digit that is more than zero, increase the previous digit by 1. When the first digit to be dropped is 4 or less, or 5 followed by a zero, do not change the previous digit.

2–2–6   Rounding Off

In many cases we will **round off** *decimal numbers because we do not need the accuracy indicated by a large number of decimal digits*. For example, it is usually unnecessary to have so many decimal digits in a value such as 74.139896428. If we were to round off this value to the nearest hundredths place, we would include two digits after the decimal point, which is 74.13. This is not accurate, however, since the digit following 74.13 was a 9 and therefore one count away from causing a reset and carry action into the hundredths column. To take into account the digit that is to be dropped when rounding off, therefore, we follow this basic 3-step procedure:

Step 1:   To round a decimal to a particular decimal place, first locate the digit in that place and call it the *rounding digit*.

Step 2:   Look at the *test digit* to the right of the rounding digit.

Step 3:   If the test digit is 5 or greater, add 1 to the rounding digit and drop all digits to the right of the rounding digit. If the test digit is less than 5, do not change the rounding digit, and drop all digits to the right of the rounding digit.

Therefore, the value 74.139896428, rounded off to the nearest hundredths place, would equal:

┌─ Rounding digit: hundredths place

$\boxed{74.13}\,9896428 = 74.14$

Test digit: First digit to be dropped is greater than 5, so rounding digit should be increased by 1.

■ **EXAMPLE**

Round off the value 74.139896428 to the nearest ten, whole number, tenth, hundredth, thousandth, and ten-thousandth.

■ *Solution:*

$\boxed{7}$ 4.139896428 rounded off to the nearest ten = 70
  └─ Test digit to be dropped is less than 5; therefore, do not change rounding digit.

$\boxed{74.}$ 139896428 rounded off to the nearest whole number = 74
  └─ Test digit to be dropped is less than 5; therefore, do not change rounding digit.

$\boxed{74.1}$ 39896428 rounded off to the nearest tenth = 74.1
  └─ Test digit to be dropped is less than 5; therefore, do not change rounding digit.

$\boxed{74.13}$ 9896428 rounded off to the nearest hundredth = 74.14
  └─ Test digit to be dropped is 5 or greater; therefore, increase rounding digit by 1 (3 to 4).

$\boxed{74.139}$ 896428 rounded off to the nearest thousandth = 74.140
  └─ Test digit to be dropped is 5 or greater; therefore, increase rounding digit by 1. Since rounding digits is 9, allow reset and carry action to occur. The steps are:
     74.1398 ← 8 carries 1 into thousandths column.
     74.130 ← 9 resets to 0, and carries 1 into hundredths column.
     74.140 ← Hundredths digit is increased by 1.

$\boxed{74.1398}$ 96428 rounded off to the nearest ten-thousandth = 74.1399
  └─ Test digit to be dropped is 5 or greater; therefore, increase rounding digit by 1.

CALCULATOR KEYS

Name: Round function
round (*value*[,#*decimals*])

Function: Value is rounded to the number of decimals specified.

Example:
round (π, 4)
      3.1416

2–2–7 Significant Places

The number of significant places describes how many digits are in the value and how many digits in the value are accurate after rounding off. For example, a number such as 347.63 is a five-significant-place (or significant-figure) value because it has five digits in five columns. If we were to round off this value to a whole number, we would get 348.00. This value would still be a five-significant-place number; however, it would now be accurate to only three significant places. Let us examine a few problems to practice using the terms *significant places* and *accurate to significant places*.

■ EXAMPLE

On a calculator, the value of π will come up as 3.141592654.

  a. Write the number π to six significant places.
  b. Give π to five significant places, and also round off π to ten-thousandths.
  c. 3.14159000 is the value of π to _____ significant places; however, it is accurate to only _____ significant places.

■ Solution:

  a. The value π to six significant places or figures is $\boxed{3.14159}$ 27

               3.14159
               ↑
                   Uses six digits or columns.

  b. The value π to five significant places.
    $\boxed{3.1415}$ 92 rounded off to ten-thousandths = 3.1416
  c. 3.14159000 is the value of π to nine significant places (has nine digits); however, it is accurate to only six significant places (because the zeros to the right are just extra, and the value 3.14159 has only six digits).

2-2-8 Percentages

Percent
In the hundred; of each hundred.

The percent sign (%) means hundredths, which as a proper fraction is $\frac{1}{100}$ or as a decimal fraction is 0.01. For example, 50% means 50 hundredths $\left(\frac{50 \div 50}{100 \div 50} = \frac{1}{2}\right)$, which is one-half. In decimal, 50% means 50 hundredths ($50 \times 0.01 = 0.5$), which is also one-half. The following shows how some of the more frequently used percentages can be expressed in decimal.

$1\% = 1 \times 0.01 = 0.01$ (For example, 1% is expressed mathematically in decimal as 0.01.)

$5\% = 5 \times 0.01 = 0.05$
$10\% = 10 \times 0.01 = 0.10$
$25\% = 25 \times 0.01 = 0.25$
$50\% = 50 \times 0.01 = 0.50$
$75\% = 75 \times 0.01 = 0.75$

Most calculators have a percent key that automatically makes this conversion to decimal hundredths. It operates as follows.

CALCULATOR KEYS

Name: Percent key

Function: Converts the displayed number from a percentage to a decimal fraction.

Example: 18.6% = ?

Press keys: [1] [8] [.] [6] [%]

Display shows: 0.186

Now that we understand that the percent sign stands for hundredths, what does the following question actually mean: What is 50% of 12? We now know that 50% is 50 hundredths $\left(\frac{50}{100}\right)$, which is one-half. This question is actually asking: What is half of 12? Expressed mathematically as both proper fractions and decimal fractions, the question would appear as follows:

Proper Fractions   **Decimal Fractions**
50% of 12 = ?   50% of 12 = ?

$= \frac{50}{100} \times 12$ $= (50 \times 0.01) \times 12$

$= \frac{1}{2} \times 12$ $= 0.5 \times 12$

$= 6$ $= 6$

Explaining this another way, we could say that percentages are fractions in hundredths. In the preceding example, therefore, the question is actually asking us to divide 12 into 100 parts and then to determine what value we would have if we had 50 of those 100 parts. Therefore, if we were to split 12 into 100 parts, each part would have a value of 0.12 ($12 \div 100 = 0.12$). Having 50 of these 100 parts would give us a total of $50 \times 0.12 = 6$.

■ TECHNICAL TRADE APPLICATION: RETAIL

If sales tax is 8% and the merchandise price is $256.00, what is the tax on the value, and what will be the total price?

■ *Solution:*

First, determine the amount of tax to be paid.

$$8\% \text{ of } \$256 =$$
$$0.08 \times 256 = \$20.48$$

Calculator sequence: $\boxed{8}\;\boxed{\%}\;\boxed{\times}\;\boxed{2}\;\boxed{5}\;\boxed{6}\;\boxed{=}$

Answer: 20.48
The total price paid will therefore equal

$$\$256.00 + \$20.48 = \$276.48$$

■ TECHNICAL TRADE APPLICATION: PERSONAL FINANCE

Which is the better buy—a $599.95 refrigerator at 25% off or a $499.95 refrigerator with a $50 manufacturer's rebate?

■ *Solution:*

$$25\% \text{ of } \$599.95 = \$150$$
$$\text{Refrigerator A: } \$599.95 - \$150.00 = \$449.95$$
$$\text{Refrigerator B: } \$499.95 - \$50.00 = \$449.95$$

The refrigerators are the same price.

2-2-9 Averages

A *mean* **average** *is a value that summarizes a set of unequal values.* This value is equal to the sum of all the values divided by the number of values. For example, if a team has scores of 5, 10, and 15, what is their average score? The answer is obtained by adding all the scores (5 + 10 + 15 = 30) and then dividing the result by the number of scores (30 ÷ 3 = 10).

Average
A single value that summarizes or represents the general significance of a set of unequal values.

■ TECHNICAL TRADE APPLICATION: ELECTRICAL

The voltage at the wall outlet in your home was measured at different times in the day and equaled 108.6, 110.4, 115.5, and 123.6 volts (V). Find the average voltage.

■ *Solution:*

Add all the voltages.

$$108.6 + 110.4 + 115.5 + 123.6 = 458.1 \text{ V}$$

Divide the result by the number of readings.

$$\frac{458.1 \text{ V}}{4} = 114.525 \text{ V}$$

The average voltage present at your home was therefore 114.525 volts.

Statistics
A branch of mathematics dealing with the collection, analysis, interpretation, and presentation of masses of numerical data.

Statistics are generally averages and can misrepresent the actual conditions. For example, if five people earn an average salary of $204,600 a year, one would believe that all five are very well paid. Studying the individual figures, however, we find out that their yearly earnings are $8000, $6000, $4000, $20,000, and $985,000 per year. Therefore,

$$8000 + 6000 + 4000 + 20{,}000 + 985{,}000 = 1{,}023{,}000$$

$$\frac{1{,}023{,}000}{5} = 204{,}600$$

If one value is very different from the others in a group, this value will heavily influence the result, creating an average that misrepresents the actual situation.

■ INTEGRATED MATH APPLICATION: SPORTS

A football player's "average number of yards gained per carry" ranking is calculated by dividing the total number of yards gained by the total number of times the player carried the football. Which of the following players had the higher average number of yards gained per carry?

Player	Total Yards	Total Carries
Payton	16,726	3,838
Brown	12,312	2,359

■ *Solution:*

$$\text{Payton: } \frac{16{,}726}{3{,}838} = 4.4 \text{ yards per carry average}$$

$$\text{Brown: } \frac{12{,}312}{2{,}359} = 5.2 \text{ yards per carry average}$$

On average, Brown gained a higher number of yards per carry.

SELF-TEST EVALUATION POINT FOR SECTION 2–2

Now that you have completed this section, you should be able to:

■ **Objective 6.** Describe the relationships among the four basic arithmetic operations: addition, subtraction, multiplication, and division.

■ **Objective 7.** Demonstrate how to perform the following arithmetic operations: decimal addition, decimal subtraction, decimal multiplication, and decimal division.

■ **Objective 8.** Describe the following mathematical terms and their associated operations: ratios, rounding off, significant places, percentages, and averages.

Use the following questions to test your understanding of Section 2–2.

1. The two basic arithmetic operations are _____ and _____.

2. Give the opposite of the following arithmetic operations.
   a. Addition
   b. Subtraction
   c. Multiplication
   d. Division

3. Perform the following arithmetic operations by hand, and then confirm your answers with a calculator.
   a. $26.443 + 197.1 + 2.1103 + 0.004 = ?$
   b. $19{,}637.224 - 866.43 = ?$
   c. $894.357 \times 8.6 = ?$
   d. $1.3397 \div 0.015 = ?$

4. Express a 176 meter to 8 meter ratio as a fraction, using the ratio sign and as a decimal.

5. Round off the following values to the nearest hundredth.
   a. 86.43760
   b. 12,263,415.00510
   c. 0.176600

6. Referring to the values in Question 5, describe:
   a. Their number of significant places
   b. To how many significant places they are accurate

7. Calculate the following.
   a. 15% of 0.5 = ?   c. 2.35% of 10 = ?
   b. 22% of 1000 = ?   d. 96% of 20 = ?

8. Calculate the average of the following values.
   a. 20, 30, 40, and 50
   b. 4000, 4010, 4008, and 3998

Use the following technical trade questions to test your understanding of practical applications of Section 2–2.

1. **Electrical** The National Fire Protection Association developed a set of standards known as the American Wire Gauge (AWG) for all copper conductors, which lists their diameter, resistance, and maximum safe current in amperes. The table lists some frequently used residential wires.
   a. Measured with a micrometer, the diameter of a wire is 0.092 inch. What is the AWG?
   b. What AWG wire size would you choose for a circuit that can draw a maximum of 21 amps?
   c. What is the diameter difference between an AWG #10 and a #12 wire?

AWG #	DIAMETER	MAX. CURRENT
10	0.102 in.	30 A
11	0.091 in.	25 A
12	0.081 in.	20 A
14	0.064 in.	15 A
15	0.051 in.	6 A

2. **Metalworking** The accompanying table lists a section of the U.S. Standard sheet metal gauge.

GAUGE #	THICKNESS
7–0	0.5 in.
6–0	0.469 in.
5–0	0.438 in.
4–0	0.406 in.
3–0	0.375 in.
2–0	0.344 in.
0	0.313 in.
1	0.281 in.
2	0.266 in.
3	0.25 in.

   a. What is the difference in thickness between gauge numbers 5–0 and 2?
   b. What would be the thickness if seven gauge number 4–0 sheets were stacked together?
   c. Which gauge number is twice as thick as gauge #3?

3. **Electronics** A 330 ohm (330 Ω) resistor is listed as having a 10% (plus or minus ten percent) tolerance, which means the manufacturer guarantees the value will not be less than 330 ohms plus 10%, or less than 330 ohms minus 10%. List the resistor's possible range of values.

4. **Masonry** About 3% to 4% of the bricks delivered a construction site cannot be used because they are too badly chipped or broken. If 1700 bricks are delivered, what range can you expect to discard?

5. **Machining** The outside diameter of a machined part is measured with a vernier caliper, and the measurement is always slightly different. First, round the values to the nearest thousandth of an inch, and then determine the average, and round your answer to the nearest thousandth of an inch.

   $$1.6542, \quad 1.6650, \quad 1.6552, \quad 1.6562$$

6. **Administration** An employee time sheet shows the following:

   Monday–$6\frac{1}{4}$ hours
   Tuesday–8 hours, 30 minutes
   Wednesday–4.5 hours
   Thursday–6 hours and $\frac{3}{4}$
   Friday–2 hours

   Convert all to decimal, and then calculate the average number of hours worked in a day.

REVIEW QUESTIONS

Multiple Choice Questions

1. The decimal point is used to:
   a. Separate the whole numbers from the fraction
   b. Separate the tens column from the units column
   c. Identify the decimal number system
   d. All the above

2. What is the positional weight of the second column to the right of the decimal point?
   a. Tenths
   b. Tens
   c. Units
   d. Hundredths

3. The Greek letter pi represents a constant value that describes how much bigger a circle's _____ is compared with its _____, and it is frequently used in many formulas involved with the analysis of circular motion.
   a. Circumference, radius
   b. Circumference, diameter
   c. Diameter, radius
   d. Diameter, circumference

4. _____ is the opposite arithmetic operation of _____, and _____ is the opposite arithmetic operation of _____.
   a. Multiplication, division, subtraction, addition
   b. Division, addition, multiplication, subtraction
   c. $+, \div, -, \times$
   d. $\times, -, \div, +$

5. Is π a ratio?
   a. Yes
   b. No

6. Round off the following number to the nearest hundredth: 74.8552.
   a. 74.85
   b. 74.86
   c. 74.84
   d. 74.855

7. The number 27.0003 is a _____ significant-place number and is accurate to _____ significant places.
   a. 6, 2
   b. 2, 6
   c. 6, 6
   d. 4, 6

8. Express 29.63% mathematically as a decimal.
   a. 29.63
   b. 2.963
   c. 0.02963
   d. 0.2963

Communication Skill Questions

9. How are whole numbers and fractions represented in decimal? (2–1)
10. Describe how written fractions are converted to decimal fractions and, conversely, how decimal fractions are converted to written fractions. (2–1)
11. How do you obtain the reciprocal of a number, and what is its relationship to the original number? (2–1)
12. Arbitrarily choose values, and describe the steps involved in: (2–1)
    a. Decimal addition
    b. Decimal subtraction
    c. Decimal multiplication
    d. Decimal division

13. What is a ratio, and how is it expressed? (2–2)
14. Briefly describe the rules for rounding off. (2–2)
15. What is a percentage? (2–2)
16. How is an average calculated? (2–2)
17. What is the approximate value of the constant pi (π) and what does it describe? (2–1)

Practice Problems

18. What are the individual digit weights of 2.7463?
19. Write each of the following values as a number.
    a. Two units and three-tenths
    b. Five-tenths and seven-thousandths
    c. Nine thousand, three tens, and four-hundredths
20. Indicate which zeros can be dropped in the following values (excluding the zero in the units place).
    a. 02.017
    b. 375.011020
    c. 0.0102000
21. Convert the following written fractions to their decimal equivalents.
    a. $\frac{16}{32}$
    b. $3\frac{8}{9}$
    c. $4\frac{9}{8}$
    d. $195\frac{7}{3}$
22. Calculate the reciprocal of the following values.
    a. π
    b. 0.707
    c. 0.3183
23. Convert the following decimal fractions to written fractions.
    a. $0.777 = \frac{?}{9}$
    b. $0.6149069 = \frac{?}{161}$
    c. $43.125 = \frac{?}{8}$

24. Perform the following arithmetic operations by hand, and then confirm your answers with a calculator.
    a. 764 + 37 + 2199 = ?
    b. 22,763 + 4 + 56,003 = ?
    c. 451 + 19 + 17 = ?
    d. 5555 + 3333 + 9999 = ?
    e. 83.57 + 4 + 0.663 = ?
    f. 898 − 457 = ?
    g. 7,660,232 − 23,000 = ?
    h. 4.9 − 3.7 = ?
    i. 22 − 14.7 = ?
    j. 2.01 − 1.1 = ?
    k. 4 × 16 = ?
    l. 488 × 14.0 = ?
    m. 0.3 × 99 = ?
    n. 22,000 × 14,000 = ?
    o. 0.5 × 10,000 = ?
    p. 488 ÷ 4 = ?
    q. 64.8 ÷ 8.1 = ?
    r. 99,000 ÷ 10 = ?
    s. 123 ÷ 17.5 = ?
    t. 0.662 ÷ 0.002 = ?

25. Express the ratio of the following in their lowest terms. (Remember to compare like quantities.)
    a. The ratio of 20 feet to 5 feet
    b. The ratio of $2\frac{1}{2}$ minutes to 30 seconds

26. Round off as indicated.
    a. 48.36 to the nearest whole number
    b. 156.3625 to the nearest tenth
    c. 0.9254 to the nearest hundredth

27. Light travels in a vacuum at a speed of 186,282.3970 miles per second. This _____-significant-place value is accurate to _____ significant places.

28. Calculate the following percentages.
    a. 23% of 50 = ?
    b. 78% of 10 = ?
    c. 3% of 1.5 = ?
    d. 20% of 3300 = ?
    e. 5% of 10,000 = ?

Math Application Practice Problems

30. **Health Care** The low-power laser used in a laser vision correction system removes 39 millionths of an inch of tissue in 12 billionths of a second. Write these two values as decimal fractions.

31. **Automotive** The vehicle specification calls for the gap in the engine spark plugs to be between 0.032 and 0.036 inch. Which of the following plugs are outside this spec., and should be replaced?

    Plug 1: 0.035″   Plug 2: 0.030″
    Plug 3: 0.029″   Plug 4: 0.035″
    Plug 5: 0.034″   Plug 6: 0.033″

32. **Electronics** The speed of light in a vacuum is 186,282.397 miles per second. Round this number:
    a. To the nearest thousand miles per second.
    b. To the nearest ten thousand miles per second.

33. **Office Management** An employee is paid mileage of 25 cents per mile. Determine what mileage should be paid for the months February through May.

MONTH	MILES DRIVEN
February	2,245
March	1,673
April	1,807
May	950

34. **Accounting** How many cells are there in an 8-column by 15-row spreadsheet?

35. **Health Care** How many pills should a pharmacist put in a container for a patient who should take two tablets, three times a day, for 16 days?

36. **Energy Management** Determine the average number of therms of natural gas used per month by the customer listed in the table.

29. Calculate the average value of the following groups of values (include units).
    a. 4, 5, 6, 7, and 8
    b. 15 seconds, 20 seconds, 18 seconds, and 16 seconds
    c. 110 volts, 115 volts, 121 volts, 117 volts, 128 volts, and 105 volts
    d. 33 ohms, 17 ohms, 1000 ohms, and 973 ohms
    e. 150 meters, 160 meters, and 155 meters

MONTH	THERMS OF GAS USED
January	38
February	40
March	43
April	31
May	29
June	10
July	12
August	14

37. **Machining** How much must be milled off the steel rod shown, so that the collar will slip over the end of it? (Allow 0.1 cm clearance.)

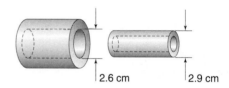

38. **Geology** Geologists classify soil types according to the grain size of the particles found in the soil. The four major classifications are shown. Classify the following soil samples:

    A (Dry Riverbed) = 0.025 in
    B (Rivers edge) = 0.00075 in
    C (River bank) = 0.00005 in
    D (Ridge) = 0.095 in

SOIL TYPE	GRAIN SIZE
Clay	0 to 0.00008 in
Silt	0.00008 in to 0.002 in
Sand	0.002 in to 0.08 in
Granule	0.08 in to 0.15 in

39. **Meteorology** When not inflated, a giant neoprene (rubber substance) weather balloon has a material thickness of 0.012 inch. When inflated with helium, the neoprene material thickness is 0.0023 inch. Determine the change in neoprene thickness.

Web Site Questions

Go to the Web site http://www.prenhall.com/cook, select the textbook *Mathematics for the Technical Trades,* select this chapter, and then follow the instructions when answering the multiple-choice practice problems.

Positive and Negative Numbers

Blaise of Genius

Blaise Pascal, the son of a regional tax official, was born in France in 1623. His father realized early that his son was a genius when his son's understanding of mathematics at the age of 8 exceeded his own. By the age of 16, Pascal had published several mathematical papers on conic sections and hydraulics. While still a youth he invented, among other things, the syringe, the hydraulic press, and the first public bus system in Paris.

At the age of 19 he began working on an adding machine in an attempt to reduce the computational drudgery of his father's job. Pascal's machine, which became known as the *Pascaline*, became the rage of European nobility and is recognized today as one of the first computers. The operator loaded a number into a machine by setting a series of wheels. Each wheel, marked with the digits zero through nine, stood for each of the decimal columns: 1s, 10s, 100s, and so on. A carry was achieved when a total greater than nine caused a complete revolution of a wheel, which advanced the wheel to the left by one digit.

Although widely praised, the Pascaline, whose adding principle of operation would remain the standard for the next 300 years, did not make Pascal rich. Like many geniuses, he was a troubled person whose ideas and problems would plague him until he could find solutions. At the age of 37, he dropped out of society to join a religious monastery that refrained from all scientific pursuits. He died just two years later at the age of 39.

In his lifetime he became famous as a mathematician, physicist, writer, and philosopher, and in honor of his achievements one of today's computer programming languages has been named after him—Pascal.

3

Outline and Objectives

VIGNETTE: BLAISE OF GENIUS

INTRODUCTION

Objective 1: Define the difference between a positive number and a negative number.

Objective 2: Describe some applications in which positive and negative numbers are used.

3–1 EXPRESSING POSITIVE AND NEGATIVE NUMBERS

Objective 3: Explain how to express positive and negative numbers.

3–2 ADDING POSITIVE AND NEGATIVE NUMBERS

Objective 4: Describe the three basic rules of positive and negative numbers as they relate to:
        a. Addition
        b. Subtraction
        c. Multiplication
        d. Division

3–3 SUBTRACTING POSITIVE AND NEGATIVE NUMBERS

3–4 MULTIPLYING POSITIVE AND NEGATIVE NUMBERS

3–5 ORDER OF OPERATIONS

3–6 DEALING WITH STRINGS OF POSITIVE AND NEGATIVE NUMBERS

Objective 5: Explain how to combine strings of positive and negative numbers.

Introduction

Positive Number
A value that is greater than zero.

Negative Number
A value that is less than zero.

Until now we have been dealing only with **positive numbers.** In fact, all the values we have discussed so far have been positive. For example, even though 12 does not have a positive (+) sign in front of it, it is still a positive number and therefore could be written as +12. We do not normally include the positive sign in front of a positive value unless we have other values that are negative. In instances when a **negative number** is present, we include the positive sign (+) in front of positive numbers to distinguish them from negative numbers that have the negative sign (−) in front of them. For example, if a list contained both positive and negative values, we would have to include their signs, as follows:

$$-4 \quad +12 \quad -6 \quad +156 \quad -198{,}765 \quad +1{,}000{,}000$$

On the other hand, if we were to list only the positive values, there would be no need to include the positive sign because all values are positive.

$$12 \quad 156 \quad 1{,}000{,}000 \quad \text{or} \quad +12 \quad +156 \quad +1{,}000{,}000$$

By definition, therefore, *a positive number is any value that is greater than zero, whereas a negative number is any value that is less than zero,* as shown in Figure 3–1.

The fractions discussed so far, which are pieces of a whole, are less than 1 and can be positive values or negative values. For example, some of the following fractions are positive (have a value between zero and positive one), and some are negative (have a value between zero and negative one).

$$-\frac{1}{2} \quad +0.5 \quad +\frac{1}{8} \quad -0.0016$$

The next important question is: *Why do we need both positive and negative numbers? The answer is: to indicate values that are above and below a reference point.* For example, consider the thermometer shown in Figure 3–2, which measures temperature in degrees Celsius (symbolized °C). This scale is used to indicate the amount of heat. On this scale, a reference point of 0 °C (zero degrees Celsius) was assigned to the temperature at which water freezes or ice melts, and this is an easy point to remember. For example, let us assume that the ice cubes inside your freezer are at a temperature of −15 °C (negative 15 degrees Celsius), and you remove several ice cubes and place them in a glass on the kitchen counter. At the warmer room temperature the ice cubes will absorb heat, and their temperature value will increase, or become less negative, as follows: −15 °C, −12 °C, −10 °C, −6 °C, and so on.

As we cross this temperature scale's reference point of 0 °C, the ice cubes change from a solid (ice cubes) to a liquid (water). If we then pour the water into a kettle and heat it even further, its temperature value will increase or become more positive, as follows: −5 °C, +12 °C, +22 °C, and so on.

Negative and positive numbers therefore are used to designate some quantity or value relative to, or compared with, a zero reference point. Temperature is not the only application for these numbers. For example, positive values can be used to express money that you have, whereas negative values can be used to express money that you owe, or have paid out. Similarly, a zero reference point can be used to indicate the average rainfall or snowfall, with positive values indicating the amount of rain or snow above the average, and negative values indicating the amount of rain or snow below the average.

In Chapter 3 we will discuss how to add, subtract, multiply, and divide positive and negative numbers, or signed numbers (numbers with signs).

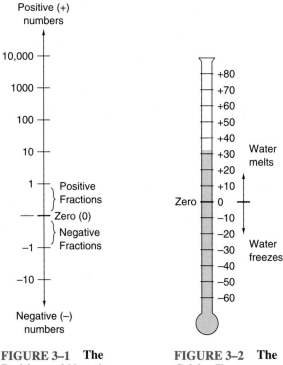

FIGURE 3–1 The Positive and Negative Number Line.

FIGURE 3–2 The Celsius Temperature Scale.

3–1 EXPRESSING POSITIVE AND NEGATIVE NUMBERS

Let us first discuss how we will mathematically express an addition, subtraction, multiplication, or division involving positive and negative numbers. As an example, let us assume that the rainfall over the last 2 years has been 3 and 9 graduations (marks on the scale) above the average reference point of zero. Because these two values of 3 and 9 are above the zero reference, they are positive numbers and can therefore be written as

$$3 \text{ and } 9 \quad \text{or} \quad +3 \text{ and } +9$$

To calculate the total amount of rainfall over the last 2 years, therefore, we must add these two values, as follows:

$$3 + 9 = 12 \quad \text{or} \quad +3 + (+9) = +12$$

If each graduation or mark on the scale was equivalent to 1 inch, over the past 2 years we would have had 12 inches of rain ($+12$) above the average of zero. Both preceding statements or expressions are identical, although the positive signs precede the values or numbers in the statement on the right. You can also see how the parentheses were used to separate the *plus sign* (indicating mathematical addition) from the *positive sign* (indicating that 9 is a positive number). If the parentheses had not been included, you might have found the statement confusing, because it would have appeared as follows:

$$+3 + +9 = +12$$

Plus sign ⤴ ⤴ Positive sign

The parentheses are included to isolate the sign of a number from a mathematical operation sign and therefore to prevent confusion.

As another example, let us imagine that last year the rainfall was 7 inches or graduations above the average, and the year before that there was a bit of a drought and the rainfall was 5 graduations or inches below the average. The total amount of rainfall for the last two years was therefore

$$+7 + (-5) = +2$$ ← Expression or meaningful combination of symbols

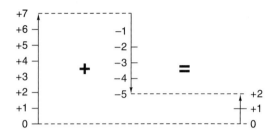   ← Graphic representation of expression

Once again, notice how the parentheses were used to separate the addition symbol (plus sign) from the sign of the number (negative 5). This example is probably best understood by referring to the graphic representation below the expression, which shows how the −5 arrow cancels 5 graduations of the +7 arrow, resulting in a sum of 2 positive graduations (+2).

Before addressing the details of adding positive and negative numbers, let us discuss the *positive/negative* or *change-sign key* available on most calculators and used to change the sign of an inputted number.

CALCULATOR KEYS

Name: Change-sign, positive/negative, or negation key

Function: Generally used after a number has been entered to change the sign of the number. Numbers with no sign indicated are always positive. Normally you press the negation key $\boxed{(-)}$ before you enter the value.

Example: $\boxed{+/-}$ $+7 + (-5) = ?$

Press keys: $\boxed{7}\,\boxed{+}\,\boxed{5}\,\boxed{+/-}\,\boxed{=}$

Answer: 2 or +2

Example: $\boxed{(-)}$ $-3 + (-15) = ?$

Press keys: $\boxed{(-)}\,\boxed{3}\,\boxed{+}\,\boxed{(-)}\,\boxed{1}\,\boxed{5}\,\boxed{=}$

Answer: −18

Note: If the subtraction key $\boxed{-}$ is used in place of the negation key $\boxed{(-)}$, or vice versa, an error occurs.

SELF-TEST EVALUATION POINT FOR SECTION 3–1

Now that you have completed this section, you should be able to:

■ **Objective 1.** Define the difference between a positive number and a negative number.

■ **Objective 2.** Describe some applications in which positive and negative numbers are used.

■ **Objective 3.** Explain how to express positive and negative numbers.

Use the following questions to test your understanding of Section 3–1.

1. Use parentheses to separate the arithmetic operation sign from the sign of the number, and then determine the result.
   a. $+3 - -4 =$
   b. $12 \times +4 =$
   c. $-5 \div -7 =$
   d. $-0.63 \times +6.4 =$

2. Express the following statements mathematically.
   a. Positive six plus negative seven
   b. Negative zero point seven five multiplied by eighteen

3. Write the calculator key sequences to input the following problems.
   a. $-7.5 + (-4.6) =$
   b. $+5 \times (-2) =$
   c. $+316.629 \div (-1.44) =$

Use the following technical trade questions to test your understanding of Section 3–1.

1. **Electrical** The direct current (dc) power supply for a notebook computer converts the 120 volt alternating current (ac) input to three dc voltage outputs.

   120 V (AC) → DC POWER SUPPLY → A: +5 V (DC), B: −7 V (DC), C: +12 V (DC)

   a. List the dc voltages present at points A, B and C.
   b. What is the difference in voltage between points A and B, B and C, A and C?

2. **Plumbing** To offset daily gasoline and equipment expenses of $65, a plumber needs to exceed this value in order to turn a profit.
   a. Express the values earned over the last two weeks as positive and negative numbers relative to $65.
   b. What is the average of the positive and negative values for week 1 and week 2?
   c. Did the plumber make a profit or stand a loss in week 1 and in week 2?

	WEEK 1	WEEK 2
M	$75	$470
T	$0	$950
W	$195	$14
T	$45	$0
F	$30	$60

3. **Automotive** An engine rod expands as it is heated and contracts as it cools. If it changes 0.4 mm per one degree Celsius (1°C), fill in the values in the list below as positive and negative numbers relative to its natural running temperature of +33°C.

   $36°C = ?$
   $35°C = ?$
   $34°C = ?$
   $+33°C = 0$
   $32°C = ?$
   $31°C = ?$
   $30°C = ?$

3–2 ADDING POSITIVE AND NEGATIVE NUMBERS

To begin, we will review how to add one positive number (+) to another positive number (+).

RULE 1: (+) + (+) = (+)

Description: A positive number (+) plus a positive number (+) equals a positive number (+).

Example:
$$5 + 15 = ?$$
$$+5 + (+15) = +20$$

Explanation:

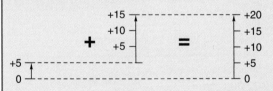

A positive 5-unit arrow plus a positive 15-unit arrow equals a positive 20-unit arrow.

Calculator sequence: [5] [+] [1] [5] [=]

Answer: 20 (+20)

Other Examples:
$$+9 + 6 = ?$$
$$+9 + (+6) = +15$$
$$81 + +16 = ?$$
$$+81 + (+16) = +97$$
$$+(15) + +15 = ?$$
$$+15 + (+15) = +30$$

The preceding box helps to collect all the facts about this rule of addition. As you know, this basic rule of addition was covered in Chapter 2.

RULE 2: $(-) + (+)$ or $(+) + (-) =$ (sign of larger number) and difference

Description: A negative number $(-)$ plus a positive number $(+)$ or a positive number $(+)$ plus a negative number $(-)$ equals the sign of the larger number and the difference.

Two Examples:

$$-8 + (+15) = ? \quad \text{or} \quad +15 + (-8) = ?$$
$$-8 + (+15) = +7 \quad\quad\quad +15 + (-8) = +7$$

Because 15 is the larger number, sign is positive.

Difference between 8 and 15 is 7.

Explanation:

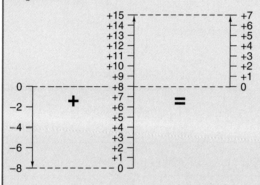

A negative 8-unit arrow plus a positive 15-unit arrow equals a positive 7-unit arrow.

Calculator sequence: $\boxed{8}\ \boxed{+/-}\ \boxed{+}\ \boxed{1}\ \boxed{5}\ \boxed{=}$

Answer: $7(+7)$

Other Examples:

$$6 + (-3) = ?$$
$$+6 + (-3) = +3 \quad \text{(larger number is } +,\ \text{difference} = 3)$$
$$-112 + +59 = ?$$
$$-112 + (+59) = -53 \quad \text{(larger number is } -,\ \text{difference} = 53)$$
$$-18 + +12 = ?$$
$$-18 + (+12) = -6 \quad \text{(larger number is } -,\ \text{difference} = 6)$$

RULE 3: $(-) + (-) = (-)$

Description: A negative number $(-)$ plus a negative number $(-)$ equals a negative number $(-)$.

Example:

$$-16 + (-4) = ?$$
$$-16 + (-4) = -20$$

Explanation:

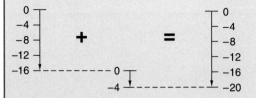

A negative 16-unit arrow plus a negative 4-unit arrow equals a negative 20-unit arrow.

Calculator sequence: [1] [6] [+/−] [+] [4] [+/−] [=]

Answer: −20

Other Examples:

$$-15 + -25 = ?$$
$$-15 + (-25) = -40$$
$$(-1063) + (-1000) = ?$$
$$-1063 + (-1000) = -2063$$

SELF-TEST EVALUATION POINT FOR SECTION 3–2

Now that you have completed this section, you should be able to:

■ *Objective 4.* *Describe the three basic rules of positive and negative numbers as they relate to addition.*

Use the following questions to test your understanding of Section 3–2.
Add the following values.

1. $+15 + (+4) = ?$
2. $-3 + (+45) = ?$
3. $+114 + (-111) = ?$
4. $-357 + (-74) = ?$
5. $17 + 15 = ?$
6. $-8 + (-177) = ?$
7. $+4600 + (-3400) = ?$
8. $-6.25 + (+0.34) = ?$

Use the following technical trade questions to test your understanding of practical applications of Section 3–2.

1. **Electrical** The current drawn by an automated manufacturing station varies based on the process being performed.

   a. What is the average current drawn compared with the zero-reference optimum load current of 6.6 amperes?

Manufacturing step	Load current
1.	−0.3 A
2.	+2.4 A
3.	+3.6 A
4.	0 A
5.	−2.2 A
6.	−1.4 A

   b. What is the actual average load current?

2. **Meteorology** The average temperature for the second week of summer should be 78 degrees Fahrenheit. The following lists the daily temperature, compared with the zero-reference point of 78 °F.

Week Day	Temperature
2/1	+2
2/2	−1
2/3	0
2/4	+6
2/5	−3
2/6	−2
2/7	−4

   a. What is the average temperature compared with the zero reference of 78 °F?
   b. What is the actual average temperature?

3–3 SUBTRACTING POSITIVE AND NEGATIVE NUMBERS

In Chapter 2 we performed a basic subtraction in the following way:

$$19 - 12 = ?$$

Steps:

$$\begin{array}{r} 19 \\ -12 \\ \hline 07 \end{array} \qquad \begin{array}{l} 9 - 2 = 7. \\ 1 - 1 = 0. \end{array}$$

Therefore, $19 - 12 = 7$.

If we wanted to, however, we could perform the same subtraction simply by changing the sign of the second number and then adding the two numbers. Let us apply this method to the previous example of $19 - 12$ and see if it works.

$19 - 12 = ?$ ← Original example
$+19 - (+12) = ?$ ← Values written as signed numbers

Step 1: Change sign of second number.
Step 2: Add the two numbers.

$+19 + (-12) = +7$ ← Same result is achieved as performing the operation $19 - 12$.

Previous rule of addition
$(+) + (-) =$ (sign of larger number) and difference

In summary, therefore, *to subtract one number from another, we can change the sign of the second number and then add the two numbers together.* If you apply this "general rule of subtraction" to all the following positive and negative number subtraction examples, you will find that it simplifies the operation because all you have to do is change the sign of the second number and then apply the rules of addition previously discussed.

RULE 1: $(+) - (+)$ or $(-) - (-) =$ (sign of larger number) and difference

Description: A positive number $(+)$ minus a positive number $(+)$, or a negative number $(-)$ minus a negative number $(-)$, equals the sign of the larger number and the difference.

Two Examples:

$$+15 - (+8) = ? \qquad\qquad -10 - (-2) = ?$$
$$+15 + (-8) = ? \qquad\qquad -10 + (+2) = ?$$

Apply rule of subtraction:
Step 1: Change sign of second number.
Step 2: Add the two numbers.

$$+15 + (-8) = +7 \qquad\qquad -10 + (+2) = -8$$

Apply rule of addition:
$(+) + (-)$ or $(-) + (+) =$ (sign of larger number) and difference

---

Explanation:

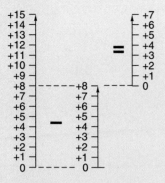

Calculator sequence: [1][5][−][8][=]   [1][0][+/−][−][2][+/−][=]

Answer: 7(+7)   −8

Other Examples:

$$-15 - (+8) = ?$$
$$-15 + (-8) = -23 \quad \leftarrow \text{(Change sign of second number and then add.)}$$

$$-2 - (-10) = ?$$
$$-2 + (+10) = +8 \quad \leftarrow \text{(Change sign of second number and then add.)}$$

$$-654 - (-601) = ?$$
$$-654 + (+601) = -53 \quad \leftarrow \text{(Change sign of second number and then add.)}$$

RULE 2: $(+) - (-) = (+)$

Description: A positive number $(+)$ minus a negative number $(-)$ will always equal a positive number $(+)$.

Example:

$$+6 - (-5) = ? \quad \longleftarrow \text{Apply rule of subtraction:}$$

Step 1: Change sign of second number.

Step 2: Add two numbers.

$$+6 + (+5) = +11 \quad \text{Apply rule of addition:}$$
$$(+) + (+) = +.$$

Explanation: As you can see from the preceding example, subtracting a negative number is the same as adding a positive number, so the result will always be positive. To explain this further, the minus sign is immediately followed by a negative sign, causing two reversals in direction, and therefore the two numbers are combined.

This *double negative* (minus sign followed by a negative sign) can also be explained by using an English language analogy. If you were to use the expression "I don't have no money," you would be using two negatives (don't and no) or reversals and therefore be saying the opposite. If you don't have "no money," you must have some money.

Calculator sequence: [6][−][5][+/−][=]

Answer: 11(+11)

Other Examples:

$$+63 - (-14) = ?$$ ← (Change sign of second number and then add.)
$$+63 + (+14) = +77$$
$$1 - {-2} = ?$$
$$+1 - (-2) = ?$$ ← (Change sign of second number and then add.)
$$+1 + (+2) = +3$$
$$(18) - {-14} = ?$$
$$+(18) - {-14} = ?$$ ← (Change sign of second number and then add.)
$$+18 + (+14) = +32$$

RULE 3: $(-) - (+) = (-)$

Description: A negative number $(-)$ minus a positive number $(+)$ will always equal a negative number $(-)$.

Example:

$$-7 - (+4) = ?$$ ← Apply rule of subtraction:

Step 1: Change sign of second number.

Step 2: Add two numbers.

$$-7 + (-4) = -11$$ Apply rule of addition:
$$(-) + (-) = -.$$

Explanation: A negative number minus a positive number $[(-) - (+) = (-)]$ is the same as a negative number plus a negative number $[(-) + (-) = (-)]$. The result will always be negative because the arithmetic operation is designed to calculate the sum of two negatives.

Calculator sequence: | 7 | +/− | − | 4 | = |

Answer: -11

Other Examples:

$$-63 - (+15) = ?$$
$$-63 + (-15) = -78$$ ← (Change sign of second number and then add.)
$$-(3) - (+15) = ?$$
$$-3 - (+15) = ?$$
$$-3 + (-15) = -18$$ ← (Change sign of second number and then add.)

SELF-TEST EVALUATION POINT FOR SECTION 3–3

Now that you have completed this section, you should be able to:

■ *Objective 4.* Describe the three basic rules of positive and negative numbers as they relate to subtraction.

Use the following questions to test your understanding of Section 3–3.

Subtract the following values.

1. $+18 - (+7) = ?$
2. $-3.4 - (-5.7) = ?$
3. $19{,}665 - (-5031) = ?$
4. $-8 - (+5) = ?$
5. $467 - 223 = ?$
6. $-331 - (-2.6) = ?$
7. $8 - (+25) = ?$
8. $-0.64 - (-0.04) = ?$

Use the following technical trade questions to test your understanding of practical applications of Section 3–3.

1. **Electronics** A quality assurance (QA) technician is checking the output voltage from a dc power supply. The +5 V (positive five volt) output has a ±0.75 V tolerance, whereas the −12 V (negative 12 volt) output has a ±2.2 V tolerance. Which of the following units should pass QA inspection? (Explain why a unit fails.)

	VOLTMETER READING OF +5 V OUTPUT	VOLTMETER READING OF +5 V OUTPUT
Unit A	+5.35 V	−11.98 V
Unit B	+4.36 V	−12.22 V
Unit C	+5.83 V	−11.75 V
Unit D	+4.29 V	−12.25 V

2. **Machining** In reference to the drilled hole, point A should be 4.2 centimeters from point *x*, and point B should be 2.6 centimeters from point *y*. If both points have a tolerance of ±0.3 cm:
   a. What is their range of acceptance?
   b. Which of the following are within or outside tolerance?

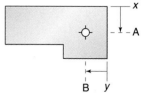

PART	1	2	3	4	5
x → A	3.8	4.2	3.9	4.6	4.2
y → B	2.4	3.0	2.5	2.6	2.7

3–4 MULTIPLYING POSITIVE AND NEGATIVE NUMBERS

Let us begin by reviewing how to multiply one positive number by another positive number. This basic rule of multiplication was covered in Chapter 2.

RULE 1: (+) × (+) = (+)

Description: A positive number (+) multiplied by a positive number (+) will always yield a positive product or result (+).

Example:

$$+6 \times (+3) = ? \quad \leftarrow (\text{same as } 6 \times 3 = ?)$$
$$+6 \times (+3) = +18$$

Explanation: Multiplication is simply repeated addition, so adding a series of positive numbers will always produce a positive result.

Six positive 3s 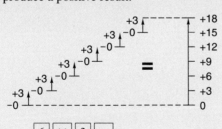 Six positive 3-unit arrows equals a positive 18-unit arrow.

Calculator sequence: [6] [×] [3] [=]

Answer: 18(+18)

Other Examples:

$$24 \times 2 = ?$$
$$+24 \times (+2) = +48$$
$$+3 \times +7 = ?$$
$$+3 \times (+7) = +21$$
$$+26{,}397 \times +31 = ?$$
$$+26{,}397 \times (+31) = +818{,}307$$

RULE 2: (+) × (−) OR (−) × (+) = (−)

Description: A positive number (+) multiplied by a negative number (−), or a negative number (−) multiplied by a positive number (+), will always yield a negative result (−).

Example:

$$+6 \times (-4) = ? \qquad \text{or} \qquad -4 \times (+6) = ?$$
$$+6 \times (-4) = -24 \qquad\qquad -4 \times (+6) = -24$$

Explanation: The result will always be negative because multiplication is simply repeated addition, and therefore adding a series of negative numbers will always result in a negative answer.

Calculator sequence: | 6 | × | 4 | +/− | = | or | 4 | +/− | × | 6 | = |
Answer: −24   −24

Other Examples:

$$18 \times (-2) = ?$$
$$+18 \times (-2) = -36$$
$$-4 \times +3 = ?$$
$$-4 \times (+3) = -12$$
$$+21.3 \times -0.2 = ?$$
$$+21.3 \times (-0.2) = -4.26$$

RULE 3: (−) × (−) = (+)

Description: A negative number (−) multiplied by a negative number (−) will always equal a positive number (+).

Example:

$$-3 \times (-5) = ?$$
$$-3 \times (-5) = +15$$

Explanation: The explanation for this is once again the double negative. After a series of negative numbers are added repeatedly, the negative answer is reversed to a positive. In summary, the reversals of direction in the operation coupled with the two negative numbers reverse the result to a positive value.

Calculator sequence: | 3 | +/− | × | 5 | +/− | = |
Answer: 15 (+15)

Other Examples:

$$-30 \times (-2) = ?$$
$$-30 \times (-2) = +60$$
$$-(2) \times -3 = ?$$
$$-2 \times (-3) = +6$$
$$-76.44 \times -1.3 = ?$$
$$-76.44 \times (-1.3) = +99.37$$

SELF-TEST EVALUATION POINT FOR SECTION 3-4

Now that you have completed this section, you should be able to:

- **Objective 4.** *Describe the three basic rules of positive and negative numbers as they relate to multiplication.*

Use the following questions to test your understanding of Section 3-4.

Multiply the following values.

1. $4 \times (+3) = ?$
2. $+17 \times (-2) = ?$
3. $-8 \times 16 = ?$
4. $-8 \times (-5) = ?$
5. $+12.6 \times 15 = ?$
6. $-3.3 \times +1.4 = ?$
7. $0.3 \times (-4) = ?$
8. $-4.6 \times -3.3 = ?$

Use the following technical trade questions to test your understanding of practical applications of Section 3-4.

1. **Electronics** The positive segment of the waveform shown on the oscilloscope display rises one-third the amplitude of the negative segment. What is the voltage of the positive peak of the waveform?

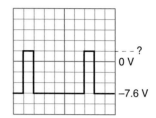

2. **Meteorology** To convert a Fahrenheit temperature to Celsius, subtract 32 from the °F reading and multiply the result by $\frac{5}{9}$. Convert the following °F readings to °C.
   a. 16 °F
   b. 76 °F
   c. 4 °F
   d. 98 °F

3-5 DIVIDING POSITIVE AND NEGATIVE NUMBERS

Let us begin by reviewing how to divide one positive number by another positive number. This basic rule of division was covered in Chapter 2.

RULE 1: $(+) \div (+) = (+)$

Description: A positive number $(+)$ divided by a positive number $(+)$ will always result in a positive number $(+)$.

Example:

$$+12 \div (+3) = ? \quad \text{(same as } 12 \div 3\text{)}$$

or

$$\frac{+12}{+3} = ?$$

$$+12 \div (+3) = +4$$

Explanation: Because division is simply repeated subtraction, subtracting a series of positive numbers from a positive number will yield a positive answer.

Calculator sequence: [1][2][÷][3][=]

Answer: 4 (+4)

Other Examples:

$$+36 \div (+2) = ?$$
$$+36 \div (+2) = +18$$

$$\frac{+504}{(+8)} = ?$$

$$+504 \div (+8) = +63$$
$$+22.7 \div +3.6 = ?$$
$$+22.7 \div (+3.6) = +6.31$$

SECTION 3-5 / DIVIDING POSITIVE AND NEGATIVE NUMBERS   75

RULE 2: (+) ÷ (−) OR (−) ÷ (+) = (−)

Description: A positive number (+) divided by a negative number (−) or a negative number (−) divided by a positive number (+) will always yield a negative result (−).

Example:

Dividend is larger than divisor:
$+16 \div (-4) = ?$ or $-16 \div (+4) = ?$
$+16 \div (-4) = -4$ or $-16 \div (+4) = -4$

Dividends (16) are larger than divisors (4).

Dividend is smaller than divisor:
$+4 \div (-16) = -0.25$ or $-4 \div (+16) = -0.25$

Even if dividends (4) are smaller than divisors (16), the answer is still negative.

Explanation: The result will always be negative because division is simply repeated subtraction, and therefore subtracting a series of negative numbers will always result in a negative answer.

Calculator sequence:

(Dividend is larger) → | 1 | 6 | ÷ | 4 | +/− | = | or | 1 | 6 | +/− | ÷ | 4 | = |

Answer: −4 −4

(Divisor is larger) → | 4 | ÷ | 1 | 6 | +/− | = | or | 4 | +/− | ÷ | 1 | 6 | = |

Answer: −0.25 −0.25

Other Examples:

$$+48 \div (-4) = ?$$
$$+48 \div (-4) = -12$$
$$-4 \div +12 = ?$$
$$-4 \div (+12) = -0.333$$
$$\frac{+677.25}{-3.66} = ?$$
$$+677.25 \div (-3.66) = -185.04$$

CHAPTER 3 / POSITIVE AND NEGATIVE NUMBERS

RULE 3: $(-) \div (-) = (+)$

Description: A negative number $(-)$ divided by a negative number $(-)$ will always equal a positive number $(+)$.

Example:

$$-15 \div (-5) = ?$$
$$-15 \div (-5) = +3$$

Explanation: The explanation is once again the double negative. After a series of negative numbers are repeatedly subtracted, the negative answer is reversed to a positive. In summary, the reversals of direction in the operation coupled with the two negative numbers reverse the result to a positive value.

Calculator sequence: [1] [5] [+/−] [÷] [5] [+/−] [=]

Answer: 3 (+3)

Other Examples:

$$-4 \div (-2) = ?$$
$$-4 \div (-2) = +2$$
$$\frac{-96}{-3} = ?$$
$$-96 \div (-3) = +32$$
$$-2.2 \div -11.0 = ?$$
$$-2.2 \div (-11.0) = +0.2$$

SELF-TEST EVALUATION POINT FOR SECTION 3–5

Now that you have completed this section, you should be able to:

■ *Objective 4.* Describe the three basic rules of positive and negative numbers as they relate to division.

Use the following questions to test your understanding of Section 3–5.

Divide the following values.

1. $\dfrac{+16.7}{+2.3} = ?$
2. $18 \div (+6) = ?$
3. $-6 \div (+2) = ?$
4. $+18 \div (-4) = ?$
5. $2 \div (-8) = ?$
6. $-8 \div +5 = ?$
7. $\dfrac{-15}{-5} = ?$
8. $0.664 \div (-0.2) = ?$

Use the following technical trade questions to test your understanding of practical applications of Section 3–5.

1. **Electronics** A -180 volt supply voltage for an oscilloscope is applied to the voltage divider shown. What will be the voltage at points A and B?

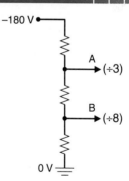

2. **Meteorology** To convert a Celsius temperature to Fahrenheit, use the following formula:

$$°F = \frac{9 \times °C}{5} + 32$$

Convert the following:
a. $-26\,°C$
b. $-8\,°C$

3–6 ORDER OF OPERATIONS

For easy access, Figure 3–3 summarizes all the rules for signed number addition, subtraction, multiplication, and division. Remember that it is more important to understand these rules rather than just to memorize them. Returning to the detailed in-chapter explanation for each of the end-of-chapter problems is not a waste of time but an investment, because you will be developing logic and reasoning skills that can be applied to all aspects of science and technology.

When a string of more than two positive and negative numbers has to be combined by some arithmetic operation, the order in which the mathematical operations should be combined is as follows:

ORDER OF OPERATIONS	MEMORY AID
First: **P**arentheses	**P**lease
Second: **E**xponents	**E**xcuse
Third: **M**ultiplication	**M**y
Fourth: **D**ivision	**D**ear
Fifth: **A**ddition	**A**unt
Sixth: **S**ubtraction	**S**ally

Order of Operations
The sequence in which a string of two or more signed numbers should be combined.

Addition	Subtraction
(+) + (+) = (+)	(+) − (+) or (−) − (−) = (sign of larger number) and difference
(−) + (+) or (+) + (−) = (sign of larger number) and difference	(+) − (−) = (+)
(−) + (−) = (−)	(−) − (+) = (−)

Multiplication	Division
(+) × (+) = (+)	(+) ÷ (+) = (+)
(+) × (−) or (−) × (+) = (−)	(+) ÷ (−) or (−) ÷ (+) = (−)
(−) × (−) = (+)	(−) ÷ (−) = (+)

FIGURE 3–3 Positive and Negative Number Rules of Addition, Subtraction, Multiplication, and Division.

This **order of operations** can more easily be remembered with the phrase "Please Excuse My Dear Aunt Sally."

■ **EXAMPLE**

Calculate the following:

a. $3 + 6 \cdot 4 = ?$

b. $4(4 + 3) = ?$

c. $8^2 - 5 = ?$

■ *Solution:*

a. Problem has multiplication and addition operations. According to the order of operations (PEMDAS), multiplication should be performed first and then addition second.

$$3 + 6 \cdot 4 = ?$$
$$3 + 24 = \qquad (6 \cdot 4 = 24)$$
$$27 \qquad (3 + 24 = 27)$$

b. $4(4 + 3) = ?$ Operations in parentheses should be performed first, multiplication second.

$$4(4 + 3) = ?$$
$$4(7) =$$
$$4 \cdot 7 =$$
$$28$$

c. $8^2 - 5 = ?$ Exponents first, subtraction second.

$$8^2 - 5 = ?$$
$$64 - 5 =$$
$$59$$

CALCULATOR KEYS

Name: Order of evaluation key

Function: Most calculators define the order in which functions in expressions are entered and evaluated. This order is as follows:

ORDER NUMBER	FUNCTION
1	Functions that precede the argument, such as $\sqrt{}($, **sin(**, or **log(**
2	Functions that are entered after the argument, such as ^{2}, $^{-1}$, $^\circ$, ^{r}, and conversions
3	Powers and roots, such as **2^5** or **5$^x\sqrt{}$32**
4	Permutations (**nPr**) and combinations (**nCr**)
5	Multiplication, implied multiplication, and division
6	Addition and subtraction
7	Relational functions, such as $>$ or $\leq$
8	Logic operator **and**
9	Logic operators **or** and **xor**

Example: Calculate $3.76 \div (-7.9 + \sqrt{5}) + 2 \log 45$.

Press keys: 3 . 76 ÷ ((−) 7 . 9 + 2nd [√]
5) + 2 × LOG (45)
ENTER

```
3.76/(−7.9 + √5)
+2 log(45)
         2.642575252
```

SELF-TEST EVALUATION POINT FOR SECTION 3–6

Now that you have completed this section, you should be able to:

■ **Objective 5.** *Explain how to combine strings of positive and negative numbers.*

Use the following questions to test your understanding of Section 3–6.

1. $+6 + (+3) + (-7) + (-5) = ?$
2. $+9 - (+2) - (-13) - (-4) = ?$
3. $-6 \times (-4) \times (-5) = ?$
4. $+8 \div (+2) \div (-5) = ?$

   or $\dfrac{\left(\dfrac{+8}{+2}\right)}{(-5)} = ?$

5. $-4 \div (-2) \times (+8) = ?$ or $\dfrac{-4}{-2} \times (+8) = ?$
6. $-9 + (+5) - (-7) = ?$

Use the following technical trade questions to test your understanding of practical applications of Section 3–6.

1. **Construction** The length the rope in feet on the drum shown is determined with the formula:

   $$L = x \times 1.05 \times y \times (x + z)$$

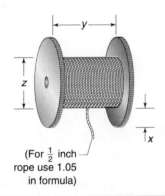

(For $\frac{1}{2}$ inch rope use 1.05 in formula)

Calculate the length of the following the order of operations if:

$x = 8$ inches

$y = 36$ inches

$z = 30$ inches

2. **Machining** A lathe cuts through a workpiece at a rate that is equal to:

   Cutting speed (feet per minute) =

   $$\frac{3.142 \times \text{diameter of workpiece} \times \text{lathe's rpm}}{12}$$

   Calculate the cutting speed if a 220 rpm lathe cuts through a 4.25 in. steel cylinder. (Indicate order of operations.)

3. **Automotive** The length of the fan belt below can be calculated with the following formula:

   $$L = ((2 \times D) + 1.57) \times (d_1 + d_2) + \frac{(d_1 + d_2)}{4 \times D}$$

   $D = 12$ inches

   $d_1 = 18$ inches

   $d_2 = 2$ inches

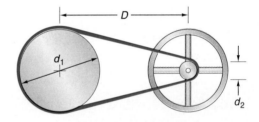

D = Distance between centers
d_1 = Diameter of load
d_2 = Diameter of driver

REVIEW QUESTIONS

Multiple Choice Questions

1. A _____ number is any value greater than zero, whereas a _____ number is any value less than zero.
   a. Positive, +
   b. −, +
   c. +, negative
   d. −, −

2. The fraction 0.3417 is a positive number.
   a. True
   b. False

3. Negative and positive numbers are used to designate some value relative to a _____ reference point.
   a. 0
   b. 0.5
   c. 10 (decimal)
   d. π

4. Which of the following calculator sequences would be used to input the following expression? $+5 - (-15) =$
   a. [5][+/−][−][1][5][=]
   b. [+/−][5][−][−][1][5][=]
   c. [5][−][−][1][5][=]
   d. [5][−][1][5][+/−][=]

5. $(+) + (+) = ?$
   a. $(+)$
   b. $(-)$
   c. (Sign of larger number) and difference
   d. None of the above

6. $(+) + (-) = ?$
   a. $(+)$
   b. $(-)$
   c. (Sign of larger number) and difference
   d. None of the above

7. $(-) + (-) = ?$
   a. $(+)$
   b. $(-)$
   c. (Sign of larger number) and difference
   d. None of the above

8. $(+) - (-)$ is equivalent to?
   a. $(-) + (-)$
   b. $(-) + (+)$
   c. $(+) + (+)$
   d. $(+) + (-)$

9. $(-) - (-)$ is equivalent to?
   a. $(-) + (+)$
   b. $(+) + (+)$
   c. $(-) + (-)$
   d. $(+) + (-)$

10. $(-) - (+)$ is equivalent to?
    a. $(-) + (+)$
    b. $(-) + (-)$
    c. $(+) + (+)$
    d. $(+) + (-)$

Communication Skill Questions

11. Define the following: (Introduction)
    a. Positive number
    b. Negative number
    c. Positive fraction
    d. Negative fraction

12. What method is used to show whether the number is positive or negative? (3–1)

13. Describe the three rules of adding positive and negative numbers. (3–2)

14. How can you achieve subtraction through addition? (3–3)

15. Describe the three rules of subtracting positive and negative numbers. (3–3)

16. What are the three rules of multiplying positive and negative numbers? (3–4)

17. List and describe the three rules of positive and negative number division. (3–5)

18. Arbitrarily choose values, and then describe the following: (3–2)
    a. $(+) + (-) = ?$
    b. $(-) + (-) = ?$
    c. $(+) + (+) = ?$

19. Arbitrarily choose values, and then describe the following: (3–3)
    a. $(+) - (+) = ?$
    b. $(-) - (+) = ?$
    c. $(+) - (-) = ?$

20. Arbitrarily choose values, and then describe the following: (3–4 and 3–5)
    a. $(+) \times (-) = ?$
    b. $(-) \times (-) = ?$
    c. $(+) \div (-) = ?$
    d. $(-) \div (-) = ?$

Practice Problems

21. Express the following statements mathematically.
  a. Positive eight minus negative five
  b. Negative zero point six times thirteen
  c. Negative twenty-two point three divided by negative seventeen
  d. Positive four divided by negative nine

22. Add the following values.
  a. $+8 + (+9) = ?$
  b. $+9 + (-6) = ?$
  c. $-6 + (+9) = ?$
  d. $-8 + (-9) = ?$

23. Subtract the following values.
  a. $+6 - (+8) = ?$
  b. $+9 - (-6) = ?$
  c. $-75 - (+62) = ?$
  d. $-39 - (-112) = ?$

24. Multiply the following values.
  a. $+18 \times (+4) = ?$
  b. $+8 \times (-2) = ?$
  c. $-20 \times (+3) = ?$
  d. $-16 \times (-10) = ?$

25. Divide the following values.
  a. $+19 \div (+3) = ?$
  b. $+36 \div (-3) = ?$
  c. $-80 \div (+5) = ?$
  d. $-44 \div (-2) = ?$

Calculate the answers to the following problems.

26. $+0.75 + (-0.25) - (+0.25) = ?$

27. $\left(\dfrac{+15}{+5}\right) \times (-3.5) = +15 \div (+5) \times (-3.5) = ?$

28. $\dfrac{+2 \times (-4) \times (-7)}{+3} = +2 \times (-4) \times (-7) \div (+3) = ?$

29. $-6 \div (-4) \div (-3) \times (-15)$

$= \dfrac{\left(\dfrac{-6}{-4}\right)}{-3} \times (-15) = ?$

30. $[+15 - (-4) - (+5) - (-10)] \div 4 = ?$

$\dfrac{+15 - (-4) - (+5) - (-10)}{4} = ?$

Math Application Practice Problems

31. Electronics Referring to the oscilloscope display shown, what is the amplitude of:
  a. Peak A
  b. Peak B
  c. Valley A
  d. Valley B

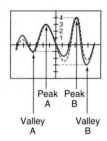

32. Automotive What is the meter reading in decimal at points Ⓐ and Ⓑ?

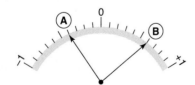

33. Water Management When drilling a well, a water table is found at 75 feet, but this soon runs dry. After an additional drill of 17 feet, an abundant water supply is discovered. Use a signed number to represent the depth of the second discovery.

34. Meteorology What is the difference between Alaska's record high and low temperatures? LOW: $-80°$, HIGH: $100°$.

35. Manufacturing New automatic insertion equipment (AIE), cuts $4\frac{1}{2}$ minutes off the production of a video processor board. If this process took 11 minutes, express the reduction in production time as a negative number.

36. Accounting A state agency predicts a budget shortfall of $6.5 million. Express this deficit as a signed number.

37. Astronomy The minimum temperature on the moon occurs just before lunar dawn, and is $-279°F$. If the difference between this temperature and the maximum temperature, which occurs at lunar noon, is $540°F$, what is the maximum temperature on the moon's surface?

38. Electrical Due to a change in load resistance, the source voltage of 240 V increases and decreases by the following during a typical day: -2 V, -4 V, $+3$ V, -6 V. What is the average change?

39. Administration Which of the following schools has the greatest shortage of computers?

HIGH SCHOOL	RATIO
Valhalla	−263
City Heights	−24
North Shore	−350
Polytech	−1010
Remenda	−63

40. Meteorology An air temperature of three degrees below zero and a wind of 27 miles per hour will give a perceived temperature of 47° below zero. Determine the change in temperature caused by the wind factor, and express it as a signed number.

Web Site Questions

Go to the Web site http://www.prenhall.com/cook, select the textbook *Mathematics for the Technical Trades,* select this chapter, and then follow the instructions when answering the multiple-choice practice problems.

Exponents and the Metric System

Not a Morning Person

René Descartes was born in 1596 in Brittany, France, and at a very early age began to display an astonishing analytical genius. At the age of 8 he was sent to Jesuit College in La Flèche, then one of the most celebrated institutions in Europe, where he studied several subjects, none of them interesting him as much as mathematics. His genius impressed all his professors, who gave him permission, because of his delicate health, to study in bed until midday—a practice that he retained throughout his life. In 1612 he left La Flèche to go to the University of Poitiers, where he graduated in law in 1616, a profession he never practiced.

Wanting to see the world, he joined the army, which made use of his mathematical ability in military engineering. In 1619 he met Dutch philosopher Isaac Beeckman, who convinced Descartes to "turn his mind back to science and worthier occupations." After leaving the army, Descartes traveled to Neuberg, Germany, where he shut himself in a well-heated room for the winter. On the eve of St. Martin's Day (November 10, 1619), Descartes described a vivid dream that determined all his future endeavors. The dream clarified his purpose and showed him that physics and all sciences could be reduced to geometry, and therefore all were interconnected "as by a chain." From this point on, his genius was displayed in his invention of coordinate geometry and in contributions to theoretical physics, methodology, and metaphysics.

In his time he was heralded as an analytical genius, a reputation that has lasted to this day. His fame was so renowned that he was asked frequently to tutor royalty. When in Paris in 1649, Descartes was asked to tutor Queen Christina of Sweden. He did not want the opportunity to, in his words, "live in the land of bears among rock and ice, and lose my independence," but he was persuaded to do so by the French ambassador a month later. The queen chose five o'clock in the morning for her lessons, and on his travels one bitter morning Descartes caught a severe chill and died within two weeks.

Descartes's problem-solving ability was incredible. You may find his four-step method to solving a problem helpful:

1. Never accept anything as true unless it is clear and distinct enough to exclude all doubt from your mind.
2. Divide the problem into as many parts as necessary to reach a solution.
3. Start with the simplest things and proceed step by step toward the complex.
4. Review the solution so completely and generally that you are sure nothing was omitted.

Outline and Objectives

VIGNETTE: NOT A MORNING PERSON

INTRODUCTION

Objective 1: Define the term *exponent*.

4–1 RAISING A BASE NUMBER TO A HIGHER POWER

Objective 2: Describe what is meant by raising a number to a higher power.

4–1–1 Square of a Number

Objective 3: Explain how to find the square and root of a number.

4–1–2 Root of a Number

4–1–3 Powers and Roots in Combination

Objective 4: Calculate the result of problems with powers and roots in combination.

4–2 POWERS OF 10

Objective 5: Explain the powers-of-10 method and how to convert to powers of 10.

4–2–1 Converting to Powers of 10

4–2–2 Powers of 10 in Combination

Objective 6: Calculate the result of problems with powers of 10 in combination.

4–2–3 Scientific and Engineering Notation

Objective 7: Describe the two following floating-point number systems:
        a. Scientific notation
        b. Engineering notation

4–3 WEIGHTS AND MEASURES

4–3–1 Customary System of Weights and Measures

Objective 8: List and describe how to convert between customary units of weight and measure.

4–3–2 Metric System of Weights and Measures

Objective 9: List the metric prefixes and describe the value of each.
    1. Metric Units of Length

Objective 10: Describe the following metric units for:
        a. Length
        b. Area
        c. Weight
        d. Volume
        e. Temperature

Objective 11: Convert U.S. customary units to metric units, and metric units to U.S. customary units.
    2. Metric Units of Area
    3. Metric Units of Weight
    4. Metric Units of Volume
    5. Metric Unit of Temperature
    6. Electrical Units, Prefixes, and Conversions

Objective 12: List and describe many of the more frequently used electrical units and prefixes, and describe how to interconvert prefixes.

4–3–3 Using the Calculator to Convert Weights and Measures

Objective 13: Show how the calculator can be used to convert between the customary and metric systems, and convert prefixes.

4–3–4 Measuring Devices

Objective 14: Describe how the micrometer and vernier calipers are used to measure the size of an object.
    1. Measuring Physical Properties
    2. Measuring Electrical Properties

Introduction

Exponent
A symbol written above and to the right of a mathematical expression to indicate the operation of raising to a power.

Like many terms used in mathematics, the word **exponent** sounds as if it will be complicated; however, once you find out that exponents are simply a sort of "math shorthand," the topic loses its intimidation. Many of the values used in science and technology contain numbers that have exponents. An exponent is a number in a smaller type size that appears to the right of and slightly higher than another number, for example:

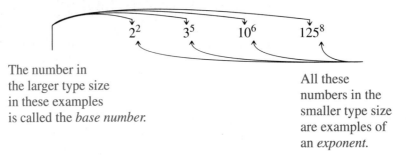

The number in the larger type size in these examples is called the *base number*.

All these numbers in the smaller type size are examples of an *exponent*.

However, what does a term like 2^3 mean? It means that the base 2 is to be used as a factor 3 times; therefore,

$$2^3 = 2 \times 2 \times 2 = 8$$

Similarly, 3^5 means that the base 3 is to be used as a factor 5 times; therefore,

$$3^5 = 3 \times 3 \times 3 \times 3 \times 3 = 243$$

As you can see, it is much easier to write 3^5 than to write $3 \times 3 \times 3 \times 3 \times 3$, and since both mean the same thing and equal the same amount (which is 243), exponents are a quick and easy math shorthand.

In this chapter we will examine the details relating to squares, roots, exponents, scientific and engineering notation, and prefixes.

4–1 RAISING A BASE NUMBER TO A HIGHER POWER

A base number's exponent indicates how many times the base number must be multiplied by itself. This is called *raising a number to a higher power*. For example, $5 \times 5 \times 5 \times 5$ can be written as 5^4, which indicates that the base number 5 is raised to the fourth power by the exponent 4. As another example, 3×3 can be written as 3^2, which indicates that the base number 3 is raised to the second power by the exponent 2. The second power is also called the **square** of the base number, and therefore 3^2 can be called "three squared" or "three to the second power."

Square
The product of a number multiplied by itself.

■ **EXAMPLE**

What does 10^3 mean?

■ *Solution:*

10^3 indicates that the base number 10 is raised to the third power by the exponent 3. Described another way, it means that the base 10 is to be used as a factor 3 times; therefore,

$$10^3 = 10 \times 10 \times 10 = 1000$$

Thus, instead of writing $10 \times 10 \times 10$, or 1000, you could simply write 10^3 (pronounced "ten to the three," "ten to the third power," or "ten cubed").

Because the square of a base number is used very frequently, let us begin by discussing raising a base number to the second power.

4-1-1 *Square of a Number*

The square of a base number means that the base number is to be multiplied by itself. For example, 4^2, which is pronounced "four squared" or "four to the second power," means 4×4. The squares of the first ten base numbers are used very frequently in numerical problems and are as follows:

$$0^2 = 0 \times 0 = 0$$
$$1^2 = 1 \times 1 = 1$$
$$2^2 = 2 \times 2 = 4$$
$$3^2 = 3 \times 3 = 9$$
$$4^2 = 4 \times 4 = 16$$
$$5^2 = 5 \times 5 = 25$$
$$6^2 = 6 \times 6 = 36$$
$$7^2 = 7 \times 7 = 49$$
$$8^2 = 8 \times 8 = 64$$
$$9^2 = 9 \times 9 = 81$$
$$10^2 = 10 \times 10 = 100$$

Many people get confused with the first three of the squares. Be careful to remember that $0^2 = 0$ because nothing × nothing equals nothing; $1^2 = 1$ because one times one is still one; and 2^2 means 2×2, not $2 + 2$, even though the answer works out both ways to be 4.

■ **EXAMPLE**

Give the square of the following.

a. 12^2 b. 7^2

■ *Solution:*

a. $12^2 = 12 \times 12 = 144$
b. $7^2 = 7 \times 7 = 49$

Finding the square of a base number is done so frequently in science and technology that most calculators have a special key just for that purpose. It is called the *square key* and operates as follows.

CALCULATOR KEYS

Name: Square key

Function: Calculates the square of the number in the display.

Example: $16^2 = ?$

Press keys: [1] [6] [x^2]

Display shows: 256

Most calculators also have a key for raising a base number of any value to any power. This is called the *y to the x power key,* and it operates as follows.

CALCULATOR KEYS

Name: *y* to the *x* power key

Function: Raises the displayed value *y* to the *x*th power.

Example: $5^4 = ?$

Press keys: [5] [y^x] [4] [=]

Display shows: 625

4–1–2   Root of a Number

What would we do if we had the result 64 and we didn't know the value of the number that was multiplied by itself to get 64? In other words, we wanted to *find the source or root number that was squared to give us the result.* This process, called **square root,** uses a special symbol called a *radical sign* and a smaller number called the *index.*

Square Root
A factor of a number that when squared gives the number.

Smaller number is called the *index.* It indicates how many times a number was multiplied by itself to get the value shown inside the radical sign. A 2 index is called the *square root.*

$\sqrt[2]{64} = ?$

Radical sign ($\sqrt{}$) indicates that the value inside is the result of a multiplication of a number two or more times.

CHAPTER 4 / EXPONENTS AND THE METRIC SYSTEM

In this example the index is 2, which indicates that the number we are trying to find was multiplied by itself two times. Of course, in this example we already know that the answer is 8 because $8 \times 8 = 64$.

If squaring a number takes us forward ($8^2 = 8 \times 8 = 64$), taking the square root of a number must take us backward ($\sqrt[2]{64} = 8$). Almost nobody extracts the squares from square root problems by hand because calculators make this process more efficient. Most calculators have a special key just for determining the square root of a number. Called the *square root key,* it operates as follows.

CALCULATOR KEYS

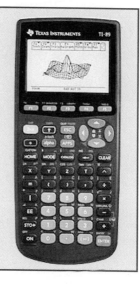

Name: Square root key

Function: Calculates the square root of the number in the display.

Example: $\sqrt{81} = ?$

Press keys: $\boxed{8}\ \boxed{1}\ \boxed{\sqrt{x}}$

Display shows: 9

Some of the more frequently used *square root* values are as follows:

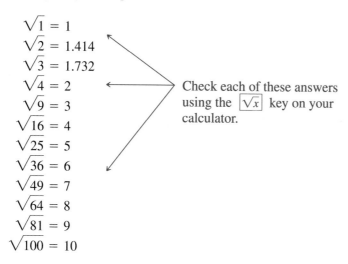

$\sqrt{1} = 1$
$\sqrt{2} = 1.414$
$\sqrt{3} = 1.732$
$\sqrt{4} = 2$
$\sqrt{9} = 3$
$\sqrt{16} = 4$
$\sqrt{25} = 5$
$\sqrt{36} = 6$
$\sqrt{49} = 7$
$\sqrt{64} = 8$
$\sqrt{81} = 9$
$\sqrt{100} = 10$

Check each of these answers using the $\boxed{\sqrt{x}}$ key on your calculator.

In some cases the index may be 3, meaning that some number was multiplied by itself three times to get the value within the radical sign; for example,

$$\sqrt[3]{125} = ?$$

Using the index 3 instead of the index 2 is called taking the **cube root.** The answer to this problem is 5 because 5 multiplied by itself three times equals 125.

$$\sqrt[3]{125} = 5$$

Cube Root
A factor of a number that when multiplied by itself three times gives the number.

Most calculators also have a key for calculating the root of any number with any index. It operates as follows.

CALCULATOR KEYS

Name: *x*th root of *y* key

Function: Calculate the *x*th root of the displayed value *y*.

Example: $\sqrt[3]{512} = ?$

Press keys: [5] [1] [2] [$\sqrt[x]{y}$] [3] [=]

Display shows: 8

In most cases a radical sign will not have an index, in which case you can assume that the index is 2, or square root; for example,

$$\sqrt{16} = 4$$

Square root of sixteen = four

4–1–3   Powers and Roots in Combination

In some instances you may have several operations to perform before the result can be determined.

■ **EXAMPLE**

What is the sum of $4^2 + 3^3$ (four squared plus three cubed)?

$$4^2 + 3^3 = ?$$

■ *Solution:*

The first step in a problem like this is to raise the base numbers to the power indicated by the exponent and then add the two values.

Steps:

$4^2 + 3^3 =$ $\qquad$ $4^2 = 4 \times 4 = 16$

$16 + 27 = 43$ $\qquad$ $3^3 = 3 \times 3 \times 3 = 27$

$\qquad\qquad\qquad\qquad$ $16 + 27 = 43$

Calculator sequence: [4] [x^2] [+] [3] [y^x] [3] [=]

Following are some other examples in which a mathematical operation needs to be performed on numbers with exponents to determine an answer or result.

■ **EXAMPLE A**

$(2 + 9)^2 = 11^2 = 121$

Exponent 2 indicates that the sum within the parentheses is to be squared.

Parentheses are used to group terms $2 + 9$.

Steps:
$2 + 9 = 11$
$11^2 = 11 \times 11 = 121$
Calculator sequence: $\boxed{2}\ \boxed{+}\ \boxed{9}\ \boxed{=}\ \boxed{x^2}$

■ **EXAMPLE B**

$(8^2)^2 = 64^2 = 4096$

Steps:
$8^2 = 8 \times 8 = 64$
$64^2 = 64 \times 64 = 4096$

Calculator sequence: $\boxed{8}\ \boxed{x^2}\ \boxed{x^2}$

■ **EXAMPLE C**

$(5^2 \times 3^3)^2 = (25 \times 27)^2$
$ = 675^2 = 455{,}625$

Steps:
$5^2 = 5 \times 5 = 25$
$3^3 = 3 \times 3 \times 3 = 27$
$25 \times 27 = 675, 675 \times 675 = 455{,}625$

Calculator sequence:
$\boxed{5}\ \boxed{x^2}\ \boxed{\times}\ \boxed{3}\ \boxed{y^x}\ \boxed{3}\ \boxed{=}\ \boxed{x^2}$

■ **EXAMPLE D**

$(5^2 - 15)^3 = (25 - 15)^3$
$ = 10^3 = 1000$

Steps:
$5^2 = 5 \times 5 = 25$
$25 - 15 = 10$
$10^3 = 10 \times 10 \times 10 = 1000$

Calculator sequence:
$\boxed{5}\ \boxed{x^2}\ \boxed{-}\ \boxed{1}\ \boxed{5}\ \boxed{=}\ \boxed{y^x}\ \boxed{3}$

■ **EXAMPLE E**

$\left(\dfrac{4^2}{2}\right)^2 + 7^2$ or $(4^2 \div 2)^2 + 7^2$
$= (16 \div 2)^2 + 7^2$
$= 8^2 + 7^2$
$= 64 + 49 = 113$

Steps:
$4^2 = 4 \times 4 = 16$
$16 \div 2 = 8$
$8^2 = 8 \times 8 = 64$
$7^2 = 7 \times 7 = 49$
$64 + 49 = 113$

Calculator sequence: $\boxed{4}\ \boxed{x^2}\ \boxed{\div}\ \boxed{2}\ \boxed{=}$

Now let us examine a few examples involving powers and roots.

EXAMPLE A

No index means square root.

$\sqrt{3 + 7 + 6} = \sqrt{16} = 4$

Radical sign extends over terms $3 + 7 + 6$.

Steps:
$3 + 7 + 6 = 16$
$\sqrt{16} = 4$

Calculator sequence: $\boxed{3}\ \boxed{+}\ \boxed{7}\ \boxed{+}\ \boxed{6}\ \boxed{=}\ \boxed{\sqrt{x}}$

EXAMPLE B

$\sqrt{14^2} = \sqrt{196} = 14$

(Because square is the opposite of square root, the square root of fourteen squared = 14.)

Steps:
$14^2 = 14 \times 14 = 196$
$\sqrt{196} = 14$

Calculator sequence: $\boxed{1}\ \boxed{4}\ \boxed{x^2}\ \boxed{\sqrt{x}}$

EXAMPLE C

$\sqrt[2]{\dfrac{8}{9}}$ or $\sqrt[2]{8 \div 9}$
$= \sqrt{0.888} = 0.943$

Steps:
$8 \div 9 = 0.889$

Calculator sequence: $\boxed{8}\ \boxed{\div}\ \boxed{9}\ \boxed{=}\ \boxed{\sqrt{x}}$

EXAMPLE D

$\sqrt[3]{5^2 \times 2^2} = \sqrt[3]{25 \times 4}$
$= \sqrt[3]{100} = 4.6416$

Steps:
$5^2 = 25$
$2^2 = 4$
$25 \times 4 = 100$
$\sqrt[3]{100} = 4.6416$

Calculator sequence: $\boxed{5}\ \boxed{x^2}\ \boxed{\times}\ \boxed{2}\ \boxed{x^2}\ \boxed{=}\ \boxed{\sqrt[x]{y}}\ \boxed{3}\ \boxed{=}$

EXAMPLE E

$\sqrt{6^2 - 5} + \sqrt{121}$
$= \sqrt{36 - 5} + \sqrt{121}$
$= \sqrt{31} + \sqrt{121}$
$= 5.568 + 11 = 16.568$

Steps:
$6^2 = 6 \times 6 = 36$
$36 - 5 = 31$
$\sqrt{31} = 5.568$
$\sqrt{121} = 11$
$5.568 + 11 = 16.568$

Calculator sequence: $\boxed{6}\ \boxed{x^2}\ \boxed{-}\ \boxed{5}\ \boxed{=}$
$\boxed{\sqrt{x}}\ \boxed{+}\ \boxed{1}\ \boxed{2}\ \boxed{1}\ \boxed{\sqrt{x}}\ \boxed{=}$

SELF-TEST EVALUATION POINT FOR SECTION 4–1

Now that you have completed this section, you should be able to:

- **Objective 1.** *Define the term* exponent.
- **Objective 2.** *Describe what is meant by raising a number to a higher power.*
- **Objective 3.** *Explain how to find the square and root of a number.*
- **Objective 4.** *Calculate the result of problems with powers and roots in combination.*

Use the following questions to test your understanding of Section 4–1.

1. Raise the following base numbers to the power indicated by the exponent, and give the answer.
   a. $16^4 = ?$
   b. $32^3 = ?$
   c. $112^2 = ?$
   d. $15^6 = ?$
   e. $2^3 = ?$
   f. $3^{12} = ?$

2. Give the following roots.
   a. $\sqrt[2]{144} = ?$
   b. $\sqrt[3]{3375} = ?$
   c. $\sqrt{20} = ?$
   d. $\sqrt[3]{9} = ?$

3. Calculate the following.
   a. $(9^2 + 14^2)^2 - \sqrt[3]{3 \times 7} = ?$
   b. $\sqrt{3^2 \div 2^2} + \dfrac{(151 - 9^2)}{3.5^2} = ?$

Use the following technical trade questions to test your understanding of practical applications of Section 4–1.

1. **Health Care** Body mass index (BMI) is used to determine whether a person is considered overweight. This value is determined with the formula:

$$BMI = \dfrac{703 \times \text{Weight (pounds)}}{\text{Height (inches)}^2}$$

20–25: Ideal
25–30: Overweight
Greater than 30: Obese

A person with a BMI of greater than 25 is considered overweight, and with a BMI greater than 30 is considered obese. What is the BMI of a 5-foot 6-inch patient who weighs 180 pounds?

2. **Law Enforcement** The maximum speed of a car involved in an accident on a wet tarmac road can be determined with the following formula:

Maximum speed =

$$\sqrt{30 \times \text{Road's coefficient of friction} \times \text{Length of skid}}$$

Calculate the maximum speed of a car based on the following data:

Road's coefficient of friction = 38% (Use 0.38.)
Length of skid = 129 feet

3. **Electronics** The total impedance, or opposition, of the series circuit shown is calculated with the formula:

$$Z \text{ (Impedance)} = \sqrt{R^2 + X^2}$$

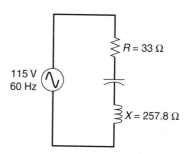

Determine Z if resistance is 33 ohms ($R = 33 \, \Omega$) and reactance is 257.8 ohms ($X = 257.8 \, \Omega$).

4. **Fire Fighting** When 65 pounds per square inch (65 psi) of pressure is present at the nozzle of a hose, the velocity of the water in feet per second (FPS) discharged is determined with the formula:

$$V(\text{FPS}) = 12.14 \sqrt{\text{pressure in psi}}$$

At what velocity will the water leave the nozzle?

4–2 POWERS OF 10

Many of the sciences deal with numbers that contain a large number of zeros, for example:

$$14,000$$

$$0.000032$$

By using exponents, we can eliminate the large number of zeros to obtain a shorthand version of the same number. This method is called *powers of 10*.

As an example, let us remove all the zeros from the number 14,000 until we are left with simply 14; however, this number (14) is not equal to the original number (14,000), and therefore simply removing the zeros is not an accurate shorthand. Another number needs to be written with the 14 to indicate what has been taken away—this number is called a *multiplier*. The multiplier must indicate what you have to multiply 14 by to get back to 14,000; therefore,

$$14,000 = 14 \times 1000$$

As you know from our discussion on exponents, we can replace the 1000 with 10^3 because $1000 = 10 \times 10 \times 10$. Therefore, the powers-of-10 notation for 14,000 is 4×10^3. To convert 14×10^3 back to its original form (14,000), simply remember that each time a number is multiplied by 10, the decimal place is moved one position to the right. In this example 14 is multiplied by 10 three times (10^3, or $10 \times 10 \times 10$), and therefore the decimal point will have to be moved three positions to the right.

$$14 \times 10^3 = 14 \times 10 \times 10 \times 10 = 14.000. = 14,000$$

As another example, what is the powers-of-10 notation for the number 0.000032? If we once again remove all the zeros to obtain the number 32, we will again have to include a multiplier with 32 to indicate what 32 has to be multiplied by to return it to its original form. In this case 32 will have to be multiplied by 1/1,000,000 (one millionth) to return it to 0.000032.

$$32 \times \frac{1}{1,000,000} = 0.000032$$

This can be verified because when you divide any number by 10, you move the decimal point one position to the left. Therefore, to divide any number by 1,000,000, you simply move the decimal point six positions to the left.

$$32 \times \frac{1}{1,000,000} = \frac{32}{1,000,000} = \frac{32}{10 \times 10 \times 10 \times 10 \times 10 \times 10} = 0.000032.$$

Once again an exponential expression can be used in place of the 1/1,000,000 multiplier, namely,

$$\frac{1}{1,000,000} = \frac{1}{10^6} = 0.000001 = 10^{-6}$$

Whenever you divide a number into 1, you get the *reciprocal* of that number. In this example, when you divide 1,000,000 into 1, you get 0.000001, which is equal to power-of-10 notation with a negative exponent of 10^{-6}. The multiplier 10^{-6} indicates that the decimal point must be moved back (to the left) by six places, and therefore

$$32 \times \frac{1}{1,000,000} = 32 \times 0.000001 = 32 \times 10^{-6} = 0.000032.$$

Now that you know exactly what a multiplier is, you have only to remember these simple rules:

1. A *negative exponent* tells you how many places *to the left* to move the decimal point.
2. A *positive exponent* tells you how many places *to the right* to move the decimal point.

CHAPTER 4 / EXPONENTS AND THE METRIC SYSTEM

Remember one important point: a negative exponent does not indicate a negative number; it simply indicates a fraction. For example, 4×10^{-3} meter means that 1 meter has been broken up into 1000 parts and we have 4 of those pieces, or 4/1000.

4-2-1 Converting to Powers of 10

To reinforce our understanding, let us try converting a few numbers to powers-of-10 notation.

■ EXAMPLE A

$$230{,}000{,}000 = 23\overset{7\,6\,5\,4\,3\,2\,1}{0{,}000{,}000} = 23 \times 10^7$$

■ EXAMPLE B

$$760{,}000 = 76\overset{4\,3\,2\,1}{0{,}000} = 76 \times 10^4$$

■ EXAMPLE C

$$0.0019 = 0.\overset{1\,2\,3\,4}{0019.} = 19 \times 10^{-4}$$

■ EXAMPLE D

$$0.00085 = 0.\overset{1\,2\,3\,4\,5}{00085.} = 85 \times 10^{-5}$$

Most calculators have a key specifically for entering powers of 10. It is called the *exponent key* and operates as follows.

CALCULATOR KEYS

Name: Exponent entry key $\boxed{\text{EXP}}$ or $\boxed{\text{EE}}$

Function: Prepares calculator to accept next digits entered as a power-of-10 exponent. The sign of the exponent can be changed by using the change-sign key $\boxed{+/-}$.

Example: 76 $\boxed{\text{EXP}}$ 4

Press keys: $\boxed{7}\,\boxed{6}\,\boxed{\text{EXP}}\,\boxed{4}$

Display shows: $\boxed{76.04}$

Example: 85×10^{-5}

Press keys: $\boxed{8}\,\boxed{5}\,\boxed{\text{EXP}}\,\boxed{5}\,\boxed{+/-}$

Display shows: $\boxed{85.\,-05}$

4–2–2 Powers of 10 in Combination

A power-of-10 multiplier raises the base number to the power of 10 indicated by the exponent. To get you used to working with powers of 10, let us do a few examples and include the calculator sequences.

■ **EXAMPLE A**

$(2 \times 10^2) + (3 \times 10^3)$
$= 200 + 3000 = 3200$

Steps:

$2 \times 10^2 = 2 \times 10 \times 10 = 200$

$3 \times 10^3 = 3 \times 10 \times 10 \times 10 = 3000$

$200 + 3000 = 3200$

Calculator sequence:

$\boxed{2}\ \boxed{\text{EXP}}\ \boxed{2}\ \boxed{+}\ \boxed{3}\ \boxed{\text{EXP}}\ \boxed{3}\ \boxed{=}$

Answer: 3200 or

$\boxed{32.02}$ (32×10^2 or 3.2×10^3)

■ **EXAMPLE B**

$(3 \times 10^6) - (2 \times 10^4)$
$= 3{,}000{,}000 - 20{,}000$
$= 2{,}980{,}000$

Steps:

$3 \times 10^6 = 3 \times 10 \times 10 \times 10 \times 10 \times 10 \times 10$
$\qquad = 3{,}000{,}000$

$2 \times 10^4 = 2 \times 10 \times 10 \times 10 \times 10$
$\qquad = 20{,}000$

$3{,}000{,}000 - 20{,}000 = 2{,}980{,}000$

Calculator sequence: $\boxed{3}\ \boxed{\text{EXP}}\ \boxed{6}\ \boxed{-}\ \boxed{2}\ \boxed{\text{EXP}}\ \boxed{4}\ \boxed{=}$

Answer: 2,980,000 or

$\boxed{298.04}$ (2.98×10^6)

■ **EXAMPLE C**

$\dfrac{1.6 \times 10^4}{4 \times 10^2}$ or

$(1.6 \times 10^4) \div (4 \times 10^2)$
$= 16{,}000 \div 400$
$= 40$

Steps:

$1.6 \times 10^4 = 1.6 \times 10 \times 10$
$\qquad\qquad \times 10 \times 10 = 16{,}000$

$4 \times 10^2 = 4 \times 10 \times 10$
$\qquad\quad = 400$

$16{,}000 \div 400 = 40$

Calculator sequence: $\boxed{1}\ \boxed{.}\ \boxed{6}\ \boxed{\text{EXP}}\ \boxed{4}\ \boxed{\div}$
$\qquad\qquad\qquad\qquad \boxed{4}\ \boxed{\text{EXP}}\ \boxed{2}\ \boxed{=}$

Answer: 40

■ **EXAMPLE D**

$(7.5 \times 10^{-6}) \times (1.86 \times 10^{-3})$
$= 0.0000075 \times 0.00186$
$= 0.00000001395$ or
1.395×10^{-8}

Steps:

$7.5 \times 10^{-6} = 0.0000075$

$1.86 \times 10^{-3} = 0.00186$

$0.0000075 \times 0.00186 = 0.00000001395$

Calculator sequence: [7] [.] [5] [EXP] [6] [+/−] [×] [1] [.] [8] [6] [EXP] [3] [+/−] [=]

Answer: 1.395×10^{-8}

4–2–3 *Scientific and Engineering Notation*

As mentioned previously, powers of 10 are used in science and technology as a shorthand owing to the large number of zeros in many values. There are basically two systems or notations used, involving values that have exponents that are a power of 10. They are called **scientific notation** and **engineering notation**.

A number in *scientific notation* is expressed as a base number between 1 and 10 multiplied by a power of 10. In the following examples, the values on the left have been converted to scientific notation.

Scientific Notation
A widely used floating-point system in which numbers are expressed as products consisting of a number between 1 and 10 multiplied by an appropriate power of 10.

Engineering Notation
A widely used floating-point system in which numbers are expressed as products consisting of a number that is greater than 1 multiplied by a power of 10 that is some multiple of 3.

■ **EXAMPLE A**

$32{,}000 = 3.2000_\circ = 3.2 \times 10^4$ — Scientific notation

Decimal point is moved to a position that results in a base number between 1 and 10. If decimal point is moved left, exponent is positive. If decimal point is moved right, exponent is negative.

■ **EXAMPLE B**

$0.0019 = 0_\circ 001.9 = 1.9 \times 10^{-3}$ — Scientific notation

■ **EXAMPLE C**

$114{,}300{,}000 = 1.14300000_\circ = 1.143 \times 10^8$ — Scientific notation

■ **EXAMPLE D**

$0.26 = 0_\circ 2.6 = 2.6 \times 10^{-1}$ — Scientific notation

Floating-Point Number System
A system in which numbers are expressed as products consisting of a number and a power-of-10 multiplier.

As you can see from the preceding examples, the decimal point floats backward and forward, which explains why scientific notation is called a **floating-point number system.** Although scientific notation is used in science and technology, the engineering notation system, discussed next, is used more frequently.

In *engineering notation* a number is represented as a base number that is greater than 1 multiplied by a power of 10 that is some multiple of 3. In the following examples, the values on the left have been converted to engineering notation.

■ EXAMPLE A

$$32,000 = 32.000 = 32 \times 10^3$$

Decimal point is moved to a position that results in a base number that is greater than 1, and a power-of-10 exponent that is some multiple of 3.

■ EXAMPLE B

$$0.0019 = 0.001.9 = 1.9 \times 10^{-3}$$

■ EXAMPLE C

$$114,300,000 = 114.300000 = 1.143 \times 10^6$$

■ EXAMPLE D

$$0.26 = 0.260. = 260 \times 10^{-3}$$

CALCULATOR KEYS

Name: Normal, scientific, engineering modes

Function: Most calculators have different notation modes that affect the way an answer is displayed on the calculator's screen. Numeric answers can be displayed with up to 10 digits and a two-digit exponent. You can enter a number in any format.

Example: Normal notation mode is the usual way we express numbers, with digits to the left and right of the decimal, as in 12345.67.

Sci (scientific) notation mode expresses numbers in two parts. The significant digits display with one digit to the left of the decimal. The appropriate power of 10 displays to the right of E, as in 1.234567E4.

Eng (engineering) notation mode is similar to scientific notation; however, the number can have one, two, or three digits before the decimal, and the power-of-10 exponent is a multiple of three, as in 12.34567E3.

SELF-TEST EVALUATION POINT FOR SECTION 4–2

Now that you have completed this section, you should be able to:

- **Objective 5.** Explain the powers-of-10 method and how to convert to powers of 10.
- **Objective 6.** Calculate the result of problems with powers of 10 in combination.
- **Objective 7.** Describe the two floating-point number systems: scientific notation and engineering notation.

Use the following questions to test your understanding of Section 4–2.

1. Convert the following to powers of 10.
   a. $100 = ?$
   b. $1 = ?$
   c. $10 = ?$
   d. $1{,}000{,}000 = ?$
   e. $\dfrac{1}{1{,}000} = ?$
   f. $\dfrac{1}{1{,}000{,}000} = ?$

2. Convert the following to common numbers without exponents.
   a. $6.3 \times 10^3 = ?$
   b. $114{,}000 \times 10^{-3} = ?$
   c. $7{,}114{,}632 \times 10^{-6} = ?$
   d. $6624 \times 10^6 = ?$

3. Perform the indicated operation on the following.
   a. $\sqrt{3 \times 10^6} = ?$
   b. $(2.6 \times 10^{-6}) - (9.7 \times 10^{-9}) = ?$
   c. $\dfrac{(4.7 \times 10^3)^2}{3.6 \times 10^6} = ?$

4. Convert the following common numbers to engineering notation.
   a. $47{,}000 = ?$
   b. $0.00000025 = ?$
   c. $250{,}000{,}000 = ?$
   d. $0.0042 = ?$

Use the following technical trade questions to test your understanding of practical applications of Section 4–2.

1. **Electrical** Convert the following measurements to their engineering notation equivalents:
   a. 98,000 watts
   b. 0.18 ampere
   c. 166,000,000 volts
   d. 3300 ohms

2. **Painting** An exterior paint covers 0.2×10^3 square feet per gallon. How many gallons would be needed to cover 6400 square feet?

3. **Plumbing** A smaller pipe will offer a larger opposition to water flow than a larger pipe. The following formula will determine how many smaller pipes are needed to supply the same flow as a larger pipe.

$$N = \sqrt{\left(\dfrac{d_L}{d_S}\right)} \quad \begin{array}{l} d_L = \text{diameter of larger pipe} \\ d_S = \text{diameter of small pipe} \end{array}$$

How many 0.75 inch $\left(\tfrac{3}{4}''\right)$ pipes are needed to supply the same flow as a 1.5 inch $\left(1\tfrac{1}{2}''\right)$ pipe?

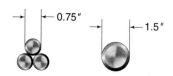

4. **Electronics** The capacity or capacitance of the capacitor shown can be calculated with the following formula:

$$C = \dfrac{(8.85 \times 10^{-12}) \times k \times A}{d}$$

where

C = Capacitance in farads (F)
k = Dielectric constant
A = Plate area in square meters (m²)
d = Distance between the plates in meters

Calculate the capacitance of the capacitor.

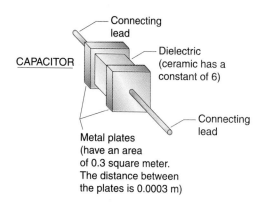

CAPACITOR — Connecting lead — Dielectric (ceramic has a constant of 6) — Connecting lead — Metal plates (have an area of 0.3 square meter. The distance between the plates is 0.0003 m)

4–3 WEIGHTS AND MEASURES

There are two systems of weights and measures used in the United States. The older **customary system of weights and measures** uses units such as the mile, square foot, and ounce, whereas the newer **metric system of weights and measures** uses units such as the meter, liter, and gram. To begin let us examine the customary system in detail.

Customary System of Weights and Measures
A non-decimal system of weights and measures based on the foot and pound system of units.

Metric System of Weights and Measures
A decimal system of weights and measures based on the meter and the gram.

4–3–1 *Customary System of Weights and Measures*

Table 4–1 lists all the most frequently used customary units for length, area, volume, and weight. Many of these are probably already familiar to you. Study the conversions and test your understanding with the following example.

TABLE 4–1   U.S. Customary System

Linear Measure (Length)
12 inches = 1 foot
3 feet = 1 yard
$5\frac{1}{2}$ yards = 1 rod
40 rods = 1 furlong
8 furlongs (5280 feet) = 1 statute mile
Therefore:
1 mile = 1760 yards = 5280 feet = 63,360 inches = 320 rods = 8 furlongs

Square Measure (Area)
144 square inches = 1 square foot
9 square feet = 1 square yard
$30\frac{1}{4}$ square yards = 1 square rod
160 square rods = 1 acre
640 acres = 1 square mile

Dry Measure (Volume)
1728 cubic inches = 1 cubic foot
27 cubic feet = 1 cubic yard
2 pints = 1 quart
8 quarts = 1 peck
4 pecks = 1 bushel

Liquid Measure (Volume)
60 minims = 1 fluid dram
8 fluid drams = 1 fluid ounce
16 fluid ounces = 1 pint
4 gills = 1 pint
2 pints = 1 quart
4 quarts = 1 gallon
$31\frac{1}{2}$ gallons = 1 barrel
2 barrels = 1 hogshead

Weight Measure
$27\frac{11}{32}$ grains = 1 dram
16 drams = 1 ounce
16 ounces = 1 pound
100 pounds = 1 short hundredweight
112 pounds = 1 long hundredweight
20 short hundredweight = 1 short ton = 2000 pounds
20 long hundredweight = 1 long ton = 2240 pounds

Mariner's Measure
6 feet = 1 fathom
1000 fathoms (approx.) = 1 nautical mile
3 nautical miles = 1 league

Surveyor's Measure
7.92 inches = 1 link
100 links = 1 chain

Wood Measure
16 cubic feet = 1 cord foot
8 cord feet = 1 cord
1 board foot (bf) = 1 square foot of lumber, 1 inch thick

Time Measure
60 seconds = 1 minute
60 minutes = 1 hour
24 hours = 1 day
7 days = 1 week
4 weeks (28 to 31 days) = 1 month
12 months (365–366 days) = 1 year
100 years = 1 century

Angular and Circular Measure
60 seconds = 1 minute
60 minutes = 1 degree
90 degrees = 1 right angle
180 degrees = 1 straight angle
360 degrees = 1 circle

■ **EXAMPLE:**

Convert the following:

a. How many feet are in a yard?
b. How many square inches are in a square foot?
c. How many pints are in a gallon?
d. How many pounds are in a ton (short)?

■ *Solution:*

a. 1 yard = 3 feet
b. 1 square foot = 144 square inches
c. 1 gallon = 8 pints
d. 1 ton (short) = 2000 pounds

To demonstrate how much easier it is to work with the metric system rather than with the customary system, let us use a simple example.

■ **EXAMPLE:**

What is 20% of:

a. 1 mile in feet (1 mile = 5280 feet)
b. 1 kilometer in meters (1 kilometer = 1000 meters)

■ *Solution:*

a. 20% of 5280 = 1056 feet
b. 20% of 1000 = 200 meters

You more than likely needed the calculator for part (a) of the example but did not need it for part (b), proving that the metric system is easier to use, to represent, to count, and to perform mathematical operations on values; however, both systems are in use, and so a good understanding of each and a knowledge of how to convert between systems is important.

4–3–2 *Metric System of Weights and Measures*

The **metric system of weights and measures** listed in Table 4–2 was developed to make working with values easier by having only a few units (meter, liter, and gram) and prefixes that are multiples of 10 (milli = 1/1000, kilo = 1000).

To help explain what we mean by units and prefixes, let us begin by examining how we measure length using the metric system. The standard **unit** of length in the metric system is the *meter* (abbreviated m). You have probably not seen the unit "meter" used a lot on its own. More

Unit
A determinate quantity adopted as a standard of measurement.

TABLE 4–2 Metric System

Linear Measure (Length)		Square Measure (Area)	
10 millimeters	= 1 centimeter	100 sq. millimeters	= 1 sq. centimeter
10 centimeters	= 1 decimeter	100 sq. centimeters	= 1 sq. decimeter
10 decimeters	= 1 meter	100 sq. decimeters	= 1 sq. meter
10 meters	= 1 decameter	100 sq. meters	= 1 sq. decameter
10 decameters	= 1 hectometer	100 sq. decameters	= 1 sq. hectometer
10 hectometers	= 1 kilometer	100 sq. hectometers	= 1 sq. kilometer
General		**General**	
1 centimeter = 10 millimeters		1 hectare (ha)	= 100 m × 100 m
1 meter = 100 cm = 1000 millimeters			= 10,000 sq. meters
1 kilometer = 1000 meters			

TABLE 4-2 (continued)

Cubic Measure (Volume)		General	
1000 cu. millimeters	= 1 cu. centimeter	1 cubic centimeter (cm^3, cc, or cu. cm) = 1 milliliter, 1000 cu. centimeters = 1 liter	
1000 cu. centimeters	= 1 cu. decimeter		
1000 cu. decimeters	= 1 cu. meter	**Weight**	
Liquid Measure (Volume)		10 milligrams	= 1 centigram
10 milliliters	= 1 centiliter	10 centigrams	= 1 decigram
10 centiliters	= 1 deciliter	10 decigrams	= 1 gram
10 deciliters	= 1 liter	10 grams	= 1 decagram
10 liters	= 1 decaliter	10 decagrams	= 1 hectogram
10 decaliters	= 1 hectoliter	10 hectograms	= 1 kilogram
10 hectoliters	= 1 kiloliter	100 kilograms	= 1 quintal
		10 quintals	= 1 ton

Prefix
An affix attached to the beginning of a word, base, or phrase.

frequently you have heard and seen the terms *centimeter, millimeter,* and *kilometer.* All these words have two parts: a *prefix name* and *unit.* For example, with the name *centimeter, centi* is the prefix, and *meter* is the unit. Similarly, with the names *millimeter* and *kilometer, milli* and *kilo* are the prefixes, and *meter* is the unit. The next question, therefore, is: What are these prefixes? A **prefix** *is simply a power of 10 or multiplier that precedes the unit.* Figure 4–1 shows the names, symbols, and values of the most frequently used metric prefixes.

What does *centimeter* mean? If you look up the prefix *centi* in Figure 4–1, you can see that it is a prefix indicating a fraction (less than one). Its value or power of 10 is 10^{-2}, or one hundredth $\left(\frac{1}{100}\right)$. This means that one meter has been divided up into 100 pieces, and each of these pieces is a centimeter or a hundredth of a meter. Millimeters have been used in photography for many years because the width of film negatives is measured in millimeters (for example, 35 mm film). Looking up the prefix *milli* in Figure 4–1, you see that its power of 10 is 10^{-3}, or one thousandth $\left(\frac{1}{1000}\right)$. This prefix is used to indicate smaller fractions because *milli* measures length in thousandths of a meter, whereas *centi* measures length in hundredths of a meter. When the length of some object is less than 1 meter, therefore, the unit *meter* will have a prefix indicating the fractional multiplier, and this power of 10 will have a negative exponent value. Remember that these negative exponents do not indicate negative numbers; they simply indicate a fractional multiplier.

Looking at the number scale in Figure 4–1, you can see that when values are between 1 and 99, we do not need to use a prefix; for example, 1 meter, 62 meters, 84 meters, and so on. A value like 4500 meters, however, would be shortened to include the $1000 (10^3)$ prefix *kilo,* as follows:

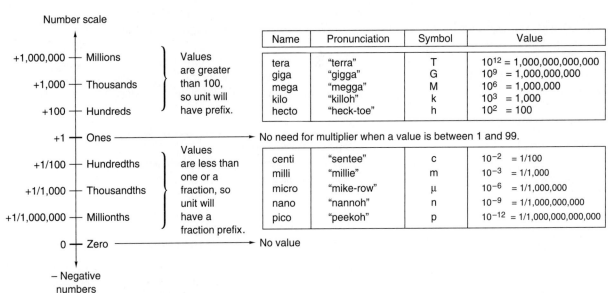

FIGURE 4–1 Metric Prefixes.

CHAPTER 4 / EXPONENTS AND THE METRIC SYSTEM

$$4500 \text{ m} = 4.5 \times 1000 \text{ m} \quad \leftarrow (4.\overset{\frown}{500}.)$$
$$= 4.5 \times 10^3 \text{ m} \quad \leftarrow \text{(Because the 1000}$$
The value 4500 $\longrightarrow$ $= 4.5 \text{ km}$ or 10^3 multiplier is *kilo*, we can substitute
is shortened to 4.5 *kilo*, or k, for 10^3.)
with the prefix kilo.

Similarly, a length of 4,600,000 meters is shortened to

$$4,600,000 \text{ m} = 4.6 \times 1,000,000 \text{ m} \quad \leftarrow (4.\overset{\frown}{600000}.)$$
$$= 4.6 \times 10^6 \text{ m} \quad \leftarrow (10^6 = \text{mega, or M})$$
$$= 4.6 \text{ Mm}$$

Now that you have an understanding of metric prefixes, let us discuss some of the metric units and compare them with our known U.S. customary units.

1. Metric Units of Length

As already mentioned, the standard unit of length in the metric system is the meter. Table 4–3 shows how to convert a metric unit of length (millimeters, centimeters, meters, kilometers, and so on) to a U.S. customary unit of length (inches, feet, yards, and miles).

TABLE 4–3 Converting Metric Units to U.S. Customary Units (Metric $\rightarrow$ U.S.)

WHEN YOU KNOW	MULTIPLY BY	TO FIND
	Length	
millimeters (mm)	0.04	inches (in.)
centimeters (cm)	0.4	inches (in.)
meters (m)	3.3	feet (ft)
meters (m)	1.1	yards (yd)
kilometers (km)	0.6	miles (mi)
	Area	
sq. centimeters (cm^2)	0.16	sq. inches (in.^{2})
sq. meters (m^2)	1.2	sq. yards (yd^2)
sq. kilometers (km^2)	0.4	sq. miles (mi^2)
hectares (ha) (10,000 m^2)	2.5	acres
	Weight	
grams (g)	0.035	ounces (oz)
kilograms (kg)	2.2	pounds (lb)
tonnes (1000 kg) (t)	1.1	short tons
	Volume	
milliliters (mL) ⎫	0.03	fluid ounces (fl oz)
liters (L) ⎬ Liquid volume	2.1	pints (pt)
liters (L)	1.06	quarts (qt)
liters (L) ⎭	0.26	gallons (gal)
cubic centimeters (cm^3) ⎫	16.387	cubic inches (in.^{3})
cubic meters (m^3) ⎬ Dry volume	35	cubic feet (ft^3)
cubic meters (m^3) ⎭	1.3	cubic yards (yd^3)
	Temperature	
Celsius (°C)	$\left(\dfrac{9}{5} \times °C\right) + 32$	Fahrenheit (°F)

■ **TECHNICAL TRADE APPLICATION: CARPENTRY**

Use Table 4–3 to find how many feet are in 1 meter of lumber.

$$1\,m = ?\,feet$$

■ *Solution:*

Referring to Table 4–3, you can see that "When You Know" meters, "Multiply by" 3.3 "To Find" feet (ft). Therefore,

$$1\,m \times 3.3 = 3.3\,ft$$

Other frequently used conversions are

$$1\,cm = 0.4\,in.$$
$$1\,km = 0.6\,mi$$

Table 4–4 reverses the process by converting U.S. customary units to metric units. For example,

$$1\,in. = 2.54\,cm$$
$$1\,ft = 30\,cm$$
$$1\,mi = 1.6\,km$$

TABLE 4–4 Converting U.S. Customary Units to Metric Units (U.S. → Metric)

WHEN YOU KNOW		MULTIPLY BY	TO FIND
Length			
inches (in.)		2.54	centimeters (cm)
feet (ft)		30	centimeters (cm)
yards (yd)		0.9	meters (m)
miles (mi)		1.6	kilometers (km)
Area			
sq. inches (in.^{2})		6.5	sq. centimeters (cm^2)
sq. feet (ft^2)		0.09	sq. meters (m^2)
sq. yards (yd^2)		0.8	sq. meters (m^2)
sq. miles (mi^2)		2.6	sq. kilometers (km^2)
acres		0.4	hectares (ha)
Weight			
ounces (oz)		28	grams (g)
pounds (lb)		0.45	kilograms (kg)
short tons (2000 lb)		0.9	tonnes (t)
Volume			
teaspoons (tsp)	⎫	5	milliliters (mL)
tablespoons (tbsp)		15	milliliters (mL)
fluid ounces (fl oz)		30	milliliters (mL)
cups (c)	⎬ Liquid volume	0.24	liters (L)
pints (pt)		0.47	liters (L)
quarts (qt)		0.95	liters (L)
gallons (gal)	⎭	3.8	liters (L)
cubic feet (ft^3)	⎱ Dry volume	0.03	cubic meters (m^3)
cubic yards (yd^3)	⎰	0.76	cubic meters (m^3)
Temperature			
Fahrenheit (°F)		$\frac{5}{9} \times (°F - 32)$	Celsius (°C)

2. Metric Units of Area

The metric unit of area is the *square meter* (abbreviated m^2). For example, an area measuring 4 meters by 3 meters contains 12 square meters (12 m^2), as shown.

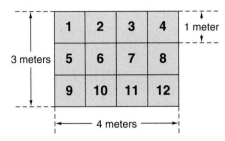

4 m by 3 m = ?

4 m × 3 m = 12 square meters or 12 m^2

■ **TECHNICAL TRADE APPLICATION: LANDSCAPING**

How many square yards are in 12 square meters of Kentucky long grass sod?

■ *Solution:*

If we wanted to find out how many square yards (yd^2) are in 12 square meters, we would use Table 4–3 (metric → U.S.).

$$12 \text{ m}^2 = ? \text{ yd}^2$$
$$12 \text{ m}^2 \times 1.2 = 14.4 \text{ yd}^2$$

3. Metric Units of Weight

In the metric system, objects are weighed in *grams*. Referring to Table 4–3, you can see that a gram is not very heavy, only 0.035 of an ounce (oz). This is why we usually measure weight in kilograms (kilo = 10^3; therefore, 1 kilogram = 1000 grams). Using Table 4–3, you can see that

$$1 \text{ kilogram (kg)} = 2.2 \text{ pounds (lb)}$$

■ **TECHNICAL TRADE APPLICATION**

A container measuring 1 foot long by 1 foot wide by 1 foot tall is filled with water and weighs 62.4 pounds. What is its metric weight in kilograms?

■ *Solution:*

$$62.4 \text{ lb} \times 0.45 = 28.1 \text{ kg}$$

4. Metric Units of Volume

Area is a space measured in two dimensions, whereas **volume** is a space measured in three dimensions, as follows:

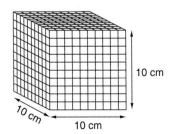

Width × height × depth =
10 cm × 10 cm × 10 cm = 1000 cm^3 or 1000 cubic centimeters

Volume
The amount of space occupied by a three-dimensional object as measured in cubic units.

The cube shown contains 1000 smaller 1 centimeter × 1 centimeter × 1 centimeter cubes. In the metric system a volume of 1000 cubic centimeters (1000 cm^3) is equal to 1 *liter* (L), which is the metric unit of *liquid volume*. Referring to Table 4–3, you can see that 1 liter is approximately equal to 1 quart (qt). This means that the next time you are at the gas station and get 2 gallons of gasoline or 8 quarts (4 quarts in 1 gallon), you have actually filled your tank with approximately 8 liters. *Dry volume* is also measured in the metric unit *meters cubed* (m^3).

5. Metric Unit of Temperature

Degrees Celsius
The international thermometer scale on which the interval between the melting point of ice (0 °C) and the boiling point of water (+100 °C) is divided into 99.9 degrees.

Degrees Fahrenheit
A thermometer scale on which the boiling point of water is at 212 degrees, and the freezing point of ice is at 32 degrees.

The metric unit of temperature is **degrees Celsius** (symbolized by °C). Most people are more familiar with the U.S. customary temperature unit, which is **degrees Fahrenheit** (symbolized by °F). These two temperature scales are compared in Figure 4–2.

As you can see by comparing the scales, the metric Celsius scale has two easy points to remember: 0 °C is the freezing or melting point of water, and 100 °C is the boiling point of water. These same points on the Fahrenheit scale are not as easy to remember (the melting or freezing point of water is 32 °F, and the boiling point of water is 212 °F).

Table 4–3 shows how to convert a °C temperature to °F, and Table 4–4 shows how to convert a °F reading to °C.

■ TECHNICAL TRADE APPLICATION: METEOROLOGY

If a thermometer reads 29 °C, what would this temperature be in degrees Fahrenheit?

■ *Solution:*

From Table 4–3, you can see that to convert a metric Celsius temperature to a U.S. customary temperature, you multiply the Celsius reading by $\frac{9}{5}$ and then add 32.

$$°F = \left(\frac{9}{5} \times °C\right) + 32$$
$$= (1.8 \times 29) + 32$$
$$= 52.2 + 32 = 84.2 \, °F$$

Steps:
$9 \div 5 = 1.82$
$1.8 \times 29 = 52.2$
$52.2 + 32 = 84.2 \, °F$

■ TECHNICAL TRADE APPLICATION: METEOROLOGY

Convert 63 °F to Celsius.

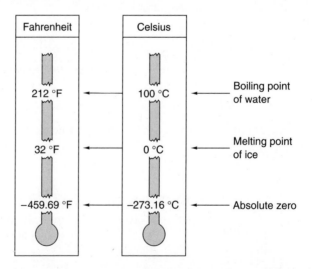

FIGURE 4–2 Comparison of Fahrenheit and Celsius Temperature Scales.

■ *Solution:*

Table 4–4 indicates that to convert °F to °C, you must subtract 32 from the Fahrenheit reading and then multiply the remainder by $\frac{5}{9}$.

$$°C = \frac{5}{9} \times (°F - 32)$$
$$= 0.55 \times (63 - 32)$$
$$= 0.55 \times 31 = 17.05 \,°C$$

Steps:
$5 \div 9 = 0.55$
$63 - 32 = 31 \,°C$

6. Electrical Units, Prefixes, and Conversions

Table 4–5 lists many of the more commonly used metric electrical quantities with their units and symbols. For example, electrical current (symbolized by I) is measured in amperes or amps (symbolized by A). As another example, electrical resistance (R) is measured in ohms (symbolized by the Greek capital letter omega, Ω).

If electrical values are much larger than one or are a fraction of one, the unit of electrical quantity is preceded by a prefix. Most electrical and electronic applications use engineering notation, so these prefixes are power-of-10 exponents that are some multiple of three (10^3, 10^6, 10^9, 10^{12}, and so on). As an example, Table 4–6 shows how these prefixes precede ampere, which is the electrical current unit. To help our understanding of these prefixes, let us do a few conversion problems.

TABLE 4–5 Metric Electrical Quantities, Their Units, and Their Unit Symbols

QUANTITY	SYMBOL	UNIT	SYMBOL
Charge	Q	coulomb	C
Current	I	ampere	A
Voltage	V	volt	V
Resistance	R	ohm	Ω
Capacitance	C	farad	F
Inductance	L	henry	H
Energy, work	W	joule	J
Power	P	watt	W
Time	t	second	s
Frequency	f	hertz	Hz

TABLE 4–6 Electrical Current Prefixes

NAME	ABBREVIATION	VALUE
Teraampere	TA	$10^{12} = 1{,}000{,}000{,}000{,}000$
Gigaampere	GA	$10^9 = 1{,}000{,}000{,}000$
Megaampere	MA	$10^6 = 1{,}000{,}000$
Kiloampere	kA	$10^3 = 1000$
Ampere	A	$10^0 = 1$
Milliampere	mA	$10^{-3} = \frac{1}{1000}$
Microampere	μA	$10^{-6} = \frac{1}{1{,}000{,}000}$
Nanoampere	nA	$10^{-9} = \frac{1}{1{,}000{,}000{,}000}$
Picoampere	pA	$10^{-12} = \frac{1}{1{,}000{,}000{,}000{,}000}$

■ **TECHNICAL TRADE APPLICATION: ELECTRONICS**

Convert the following.

    a. 0.003 A = _____ mA (milliamperes)
    b. 0.07 mA = _____ μA (microamperes)
    c. 7333 mA = _____ A (amperes)
    d. 1275 μA = _____ mA (milliamperes)

■ *Solution:*

    a. 0.003 A = _____ mA. In this example, 0.003 A has to be converted so that it is represented in milliamperes (10^{-3}, or $\frac{1}{1000}$ of an ampere). The basic algebraic rule to remember is that the expressions on either side of the equals sign must be the same.

Left		Right	
Base	Exponent	Base	Exponent
0.003 × 10^0		_____ × 10^{-3}	

    The exponent on the right in this problem will be decreased 1000 times (10^0 to 10^{-3}), so for the statement to balance, the number on the right will have to be increased 1000 times; that is, the decimal point will have to be moved to the right three places (0.003 to 3). Therefore,

$$0.003 \times 10^0 = 3 \times 10^{-3}$$

or $0.003 \text{ A} = 3 \times 10^{-3} \text{ A}$ or 3 mA

    b. 0.07 mA = _____ μA. In this example the exponent is going from milliamperes to microamperes (10^{-3} to 10^{-6}), or 1000 times smaller, so the number must be made 1000 times greater.

$$0.070. \quad \text{or} \quad 70.0$$

Therefore, 0.07 mA = 70 μA.

    c. 7333 mA = _____ A. The exponent is going from milliamperes to amperes, increasing 1000 times, so the number must decrease 1000 times.

$$7333. \quad \text{or} \quad 7.333$$

Therefore, 7333 mA = 7.333 A.

    d. 1275 μA = _____ mA. The exponent is changing from microamperes to milliamperes, an increase of 1000 times, so the number must decrease by the same factor.

$$1.2750 \quad \text{or} \quad 1.275$$

Therefore, 1275 μA = 1.275 mA.

4–3–3 *Using the Calculator to Convert Weights and Measures*

Some calculators have the ability to convert between the customary and metric systems, and to convert prefixes within a system, as shown in the following examples.

CALCULATOR KEYS

Name: Convert key ▶

Function: Some calculators have a function that converts an expression from one unit to another.

$$expression\_unit \; \blacktriangleright \; \_unit2 \Rightarrow expression\_unit2$$

Examples: How many feet are in 3 meters?

Press keys: 3_m ▶ _ft ENTER

Answer: 9.842_ft

How many kilometers are in 4 light-years?

Press keys: 4_ltyr ▶ _km

Answer: 3.78421E13_km

How many ohms are in 4 kiloohms?

Press keys: 4_kΩ ▶ _Ω

Answer: 4000_Ω

To convert a temperature value, you must use tmpCnv() instead of the ▶ operator.

Example:

Press keys: tmpCnv(100_°C to _°F)
Answer: 212_°F

4-3-4 *Measuring Devices*

Many instruments are available to measure properties. In this section we discuss the micrometer, vernier calipers, and multimeter.

1. Measuring Physical Properties

Several instruments are available for measuring the length of an object. The most basic devices are the ruler and tape measure, shown in Figure 4–3. The scale may be in customary units (inches and feet) and/or metric units (centimeters and meters). Smaller customary scales are available ($\frac{1}{32}$ and $\frac{1}{64}$ inch) when precision measurements have to be made.

The steel ruler is ideal for measuring lengths to an accuracy of $\frac{1}{32}$ or $\frac{1}{64}$ inch; however, for greater accuracy, the micrometer, shown in Figure 4–4, and the vernier calipers, shown in Figure 4–5, can be used to measure lengths to an accuracy of $\frac{1}{1000}$ inch. Follow the step-by-step procedure, and associated example, to see how these instruments are used.

MEASURING LENGTH WITH A MICROMETER	EXAMPLE (Figure 4-4)
Step 1: Read the largest number on the sleeve, which measure $\frac{1}{10}$ (0.10) inch.	$2 \times 0.100 = 0.200$ inch
Step 2: Read the additional smaller scale graduations visible on the sleeve, which measure $\frac{25}{1000}$ (0.025) inch.	$3 \times 0.025 = 0.075$ inch
Step 3: Determine the total number of graduations on the thimble scale up to the sleeve's horizontal line, which measure $\frac{1}{1000}$ (0.001) each.	$14 \times 0.001 = 0.014$ inch
Step 4: Add all of the products to determine the measurement.	0.200 0.075 +0.014 0.289 $\frac{289}{1000}$ inch

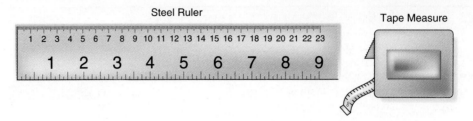

FIGURE 4–3 Basic Length Measurement Devices.

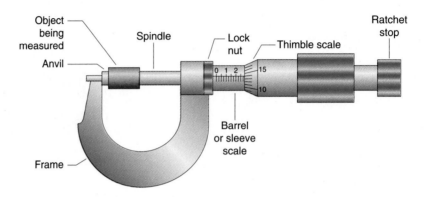

FIGURE 4–4 The Micrometer.

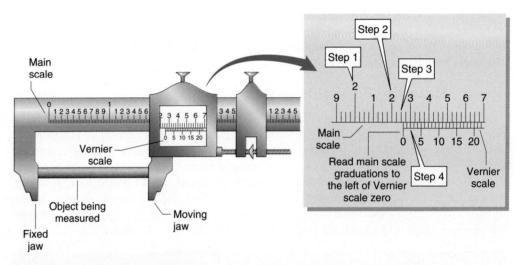

FIGURE 4–5 The Vernier Caliper.

■ EXAMPLE

Determine the length of the object being measured by the customary units micrometer shown.

■ *Solution:*

$$8 \times 0.100 = 0.800$$
$$0 \times 0.025 = 0.000$$
$$7 \times 0.001 = \underline{0.007}$$
$$0.807 \text{ inch}$$

CHAPTER 4 / EXPONENTS AND THE METRIC SYSTEM

■ **EXAMPLE:**

The metric micrometer shown is read in the same way as the customary units micrometer, except the graduations are:

Step 1:   1 mm   (one millimeter)
Step 2:   0.5 mm ($\frac{5}{10}$ millimeter)
Step 3:   0.01 mm ($\frac{1}{100}$ millimeter)

Determine the length of the object being measured by the metric micrometers shown.

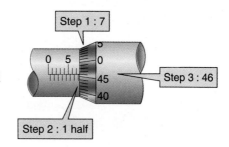

■ *Solution:*

7×1 mm $=$ 7.00
1×0.5 mm $=$ 0.50
46×0.01 mm $=$ $\underline{0.46}$
                       7.96 millimeters

MEASURING LENGTH WITH VERNIER CALIPERS	EXAMPLE (Figure 4–5)
Step 1: Read inch graduations on the main scale to the left of vernier scale zero.	$2 \times 1.000 = 2.000$ inches
Step 2: Read $\frac{1}{10}$ inch graduations visible on the main scale to the left of vernier scale zero.	$2 \times 0.100 = 0.200$ inch
Step 3: Read $\frac{25}{1000}$ inch graduations on the main scale to the left of vernier scale zero.	$2 \times 0.025 = 0.050$ inch
Step 4: Determine which is the first graduation on the vernier scale to line up exactly with any line on the main scale. Vernier scale graduations measure $\frac{1}{1000}$ inch.	2 line on vernier scale aligns with $\frac{3}{10}$ line on main scale. $2 \times 0.001 = 0.002$ inch
Step 5: Add all the products to determine the measurement.	2.000 0.200 0.050 $+0.002$ 2.252 inches

■ **EXAMPLE**

Determine the length of the object being measured by the vernier calipers shown.

■ *Solution:*

1×1.000 in. $=$ 1.000
6×0.100 in. $=$ 0.600
3×0.025 in. $=$ 0.075
5×0.001 in. $=$ $\underline{0.005}$
                          1.680 inches

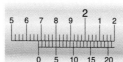

2. Measuring Electrical Properties

A multimeter is, as its name implies, a multipurpose meter. The step-by-step procedures listed in Figures 4–6, 4–7, and 4–8 show how this instrument can be used to measure electrical current, voltage, and resistance.

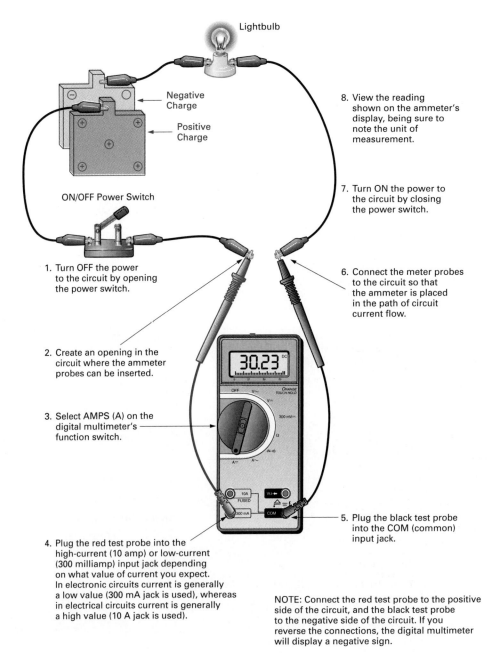

FIGURE 4–6   Measuring Current with the Ammeter.

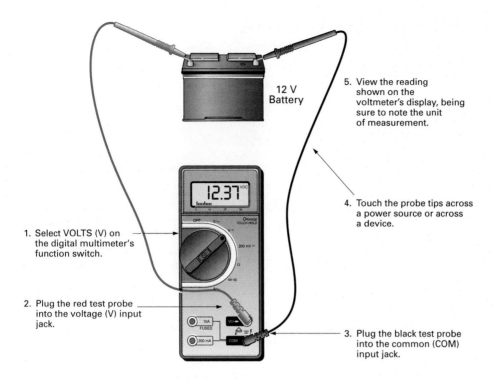

FIGURE 4–7 Measuring Voltage with the Voltmeter.

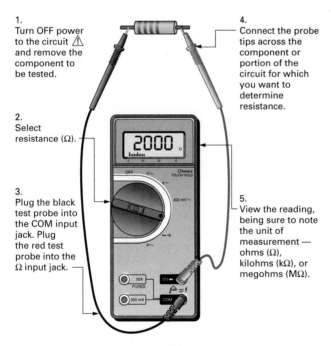

FIGURE 4–8 Measuring Resistance with the Ohmmeter.

SELF-TEST EVALUATION POINT FOR SECTION 4–3

Now that you have completed this section, you should be able to:

- **Objective 8.** List and describe how to convert between customary units of weight and measure.
- **Objective 9.** List the metric prefixes and describe the value of each.
- **Objective 10.** Describe the metric units for length, area, weight, volume, and temperature.
- **Objective 11.** Convert U.S. customary units to metric units, and metric units to U.S. customary units.
- **Objective 12.** List and describe many of the more frequently used electrical units and prefixes, and describe how to interconvert prefixes.
- **Objective 13.** Show how the calculator can be used to convert between the customary and metric systems, and to convert prefixes.
- **Objective 14.** Describe how the micrometer and vernier calipers are used to measure the size of an object.

Use the following questions to test your understanding of Section 4–3.

1. Give the power-of-10 value for the following prefixes.
   a. kilo
   b. centi
   c. milli
   d. mega
   e. micro

2. Convert the following metric units to U.S. customary units.
   a. 15 cm to inches
   b. 23 kg to pounds
   c. 37 L to gallons
   d. 23 °C to degrees Fahrenheit

3. Convert the following U.S. customary units to metric.
   a. 55 miles per hour (mph) to kilometers per hour (km/h)
   b. 16 gal to liters
   c. 3 yd^2 to square meters
   d. 92 °F to degrees Celsius

4. List the names of the metric units for the following quantities.
   a. Length
   b. Weight
   c. Temperature
   d. Time
   e. Power
   f. Current
   g. Voltage
   h. Volume (liquid)
   i. Energy
   j. Resistance

5. Convert the following.
   a. 25,000 V = _____ kilovolts
   b. 0.014 W = _____ milliwatts
   c. 0.000016 µF = _____ nanofarads
   d. 3545 kHz = _____ megahertz

Use the following technical trade questions to test your understanding of practical applications of Section 4–3.

1. **Carpentry** The construction industry measures the volume of lumber in the unit board feet. As shown, *one board foot* (1 bf) is a piece of lumber that measures one foot by one foot by one inch. To find the amount of board feet in a load of lumber, therefore, you can use the following formula:

$$N = l \times w \times h$$

where
   N = number of board feet
   l = length in feet
   w = width in feet
   h = height in inches

How many board feet are in an 8-foot 2-by-4 that measures:

$1\frac{1}{2}$ inches by $3\frac{1}{2}$ inches by 8 feet

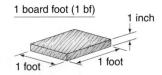

2. **Machining** What is the size of the machined part measured, if the metric micrometer is as shown?

3. **Electrical** In the illustration on the next page, which multimeter is measuring the car battery voltage and which is monitoring the current being drawn by the music system? The product of voltage and current determines power:

$$P = V \times I$$

How much power is being consumed by the music system?

114 CHAPTER 4 / EXPONENTS AND THE METRIC SYSTEM

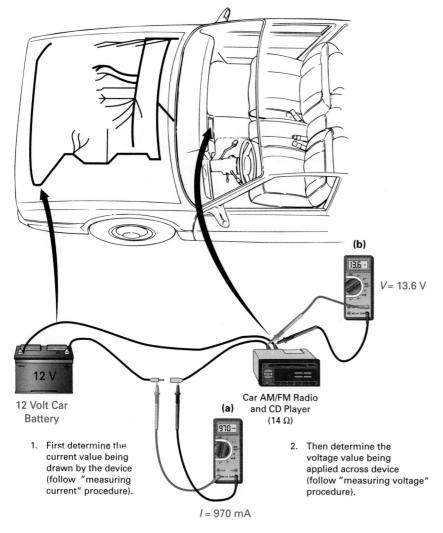

4. **Photography** A negative of the 120 film size measures $2\frac{1}{4}$ inches by $2\frac{3}{4}$ inches. What would this size be in millimeters?

5. **Automotive** An engine's cylinder displacement is measured in either cubic inches or liters. Determine the cylinder displacement of a 600 cubic inch engine in liters (round to the nearest liter).

6. **Landscaping** A cord of wood measures 8 feet long, by 4 feet wide, by 4 feet high. What would be the size of a metric cord (in centimeters and meters)?

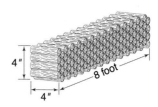

REVIEW QUESTIONS

Multiple Choice Questions

1. A base number's exponent indicates how many times the base number must be multiplied by:
   a. 2
   b. 3
   c. Itself
   d. Both (a) and (c)

2. What does 16^3 mean?
   a. 16×3
   b. $16 \times 16 \times 16$
   c. $16 \times (3 \times 3 \times 3)$
   d. $3 \times 3 \times 3 \times 3 \times 3 \times 3 \times 3 \times 3 \times 3 \times 3 \times 3 \times 3 \times 3 \times 3 \times 3 \times 3$

3. If $x^2 = 16$, then $\sqrt[2]{16} = x$. What is the value of x?
   a. 4
   b. 2
   c. 16
   d. 8

4. Calculate $\sqrt[5]{32}$.
   a. 2
   b. 1.3
   c. 4
   d. 6.4

5. Raise the following number to the power of 10 indicated by the exponent, and give the answer: 3.6×10^2.
   a. 3600
   b. 36
   c. 0.036
   d. 360

6. Convert the following number to powers of 10: 0.00029.
   a. 0.29×10^{-3}
   b. 2.9×10^{-4}
   c. 29×10^{-5}
   d. All the above

7. What is the power-of-10 value for the metric prefix *milli*?
   a. 10^{-3}
   b. 10^{-2}
   c. 10^3
   d. All the above

8. What is the metric unit for temperature?
   a. Fahrenheit
   b. Borg
   c. Celsius
   d. Rankine

9. Convert the following value of inductance to engineering notation: 0.016 henry (H).
   a. 1.6×10^2 H
   b. 16 mH
   c. 1.6 cH
   d. 16 kH

10. How many centimeters are in 8 inches?
    a. 2.03 m
    b. 3.2 cm
    c. 8 cm
    d. 20.32 cm

Communication Skill Questions

11. What is an exponent? (Introduction)
12. Define the following terms: (4–1)
    a. Square of a number
    b. Square root of a number
13. What is a power-of-10 exponent? (4–2)
14. Describe the scientific and engineering notation systems. (4–2)
15. What is the metric system? (4–3)
16. List the metric prefixes and their values. (4–3)
17. Explain how to convert the following: (4–3)
    a. Centimeters to inches
    b. Ounces to grams
    c. Yards to meters
    d. Miles to kilometers
18. What is the metric unit of: (4–3)
    a. Area
    b. Length
    c. Weight
    d. Volume
    e. Temperature
19. Give the units for the following electrical quantities: (4–3)
    a. Current
    b. Length
    c. Resistance
    d. Power
20. Arbitrarily choose values, and then describe the following conversions: (4–3)
    a. Amps to milliamps
    b. Volts to kilovolts

Practice Problems

21. Determine the square of the following values.
    a. 9^2
    b. 6^2
    c. 2^2
    d. 0^2
    e. 1^2
    f. 12^2

22. Determine the square root of the following values.
    a. $\sqrt{81}$
    b. $\sqrt{4}$
    c. $\sqrt{0}$
    d. $\sqrt{36}$
    e. $\sqrt{144}$
    f. $\sqrt{1}$

23. Raise the following base numbers to the power indicated by the exponent, and give the answer.
    a. 9^3
    b. 10^4
    c. 4^6
    d. 2.5^3
24. Determine the roots of the following values.
    a. $\sqrt[3]{343} = ?$
    b. $\sqrt{1296} = ?$
    c. $\sqrt[4]{760} = ?$
    d. $\sqrt[6]{46.656} = ?$
25. Convert the following values to powers of 10.
    a. $\dfrac{1}{100}$
    b. 1,000,000,000
    c. $\dfrac{1}{1000}$
    d. 1000
26. Convert the following to proper fractions and decimal fractions.
    a. 10^{-4}
    b. 10^{-2}
    c. 10^{-6}
    d. 10^{-3}
27. Convert the following values to powers of 10 in both scientific and engineering notation.
    a. 475
    b. 8200
    c. 0.07
    d. 0.00045
28. Convert the following to whole-number values with a metric prefix.
    a. 0.005 A
    b. 8000 m
    c. 15,000,000 Ω
    d. 0.000016 s
29. Convert the following metric units to U.S. customary units.
    a. 100 km
    b. 29 m^2
    c. 67 kg
    d. 2 L
30. Convert the following U.S. customary units to metric.
    a. 4 in.
    b. 2 mi^2
    c. 4 oz
    d. 10 gal
31. Convert the following temperatures to °C.
    a. 32 °F
    b. 72 °F
32. Convert the following temperatures to °F.
    a. 6 °C
    b. 32 °C
33. State the metric unit of:
    a. Length
    b. Area
    c. Weight
    d. Dry volume
    e. Liquid volume
    f. Temperature
34. What is the metric unit of time?
35. Convert the following.
    a. 8000 ms = _____ microseconds
    b. 0.02 MV = _____ kilovolts
    c. 10 km = _____ meters
    d. 250 mm = _____ centimeters

Math Application Practice Problems

36. **Electrical** As light travels away from its source it spreads, or diverges. At a distance of 1 yard, the light illuminates an area of 1 square unit. Expressed as an exponent, how much area is covered at 2 yards, 3 yards, and 4 yards?

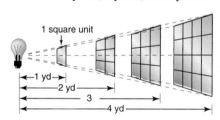

37. **Health Care** It takes an hour for a cell to divide to form another cell. In three hours, 2^3 cells exist.
    a. What is the base and what is the exponent in the exponential expression 2^3?
    b. How many cells exist after eight hours?
38. **Construction** An Olympic-size pool is 50 meters long. How many 1-foot tiles would be needed for one side of the pool?
39. **Office Management** A package weighs 32.5 kilograms. Can it be sent priority mail if the maximum limit is 70 pounds?

40. **Health Care** A sphygmomanometer measures blood pressure, and read at two points, it is expressed as a ratio such as $\frac{122}{80}$ (spoken as 122 over 80). This reading indicates that the patient has a systolic pressure of 122 millimeters of mercury, and a diastolic pressure of 80 millimeters of mercury. What would this reading be in centimeters of mercury?
41. **Painting** A painter estimates he will need 5 times as much paint as he has already used. If he has used 4 quarts, how many gallons will he need?
42. **Electronics** A handheld scopemeter measures the power consumed by a 12–volt TV/DVD system as 3286 milliwatts. How many watts is this?
43. **Health Care** A dextrose solution is being administered to a patient at a rate of 14 milliliters per hour. How many hours can a 1-liter IV bag be used?
44. **Electrical** During the month of August, a single family home uses 738 kilowat hours of electricity. What is the daily rate of electric consumption in kilowatt hours per day?
45. **Mechanical** What is the difference in horsepower between a $3\frac{1}{4}$ hp keyed-shaft motor and a $1\frac{1}{2}$ hp thru-bolt mount motor?

Web Site Questions

Go to the Web site http://www.prenhall.com/cook, select the textbook *Mathematics for the Technical Trades,* select this chapter, and then follow the instructions when answering the multiple-choice practice problems.

Algebra, Equations, and Formulas

Back to the Future

Each part of a computer system is designed to perform a specific task. You would think that all these units were first thought of by some recent pioneer in the twentieth century, but in fact, two of these elements were first described in 1833.

Born in England in 1791, Charles Babbage became very well known for both his mathematical genius and eccentric personality. For many years Babbage occupied the Cambridge chair of mathematics, once held by Isaac Newton, and although he never delivered a single lecture, he wrote several papers on numerous subjects, ranging from politics to manufacturing techniques. He also helped develop several practical devices, including the tachometer and the railroad cowcatcher.

Babbage's ultimate pursuit, however, was that of mathematical accuracy. He delighted in spotting errors in everything from log tables (used by astronomers, mathematicians, and navigators) to poetry. In fact, he once wrote to poet Alfred Lord Tennyson, pointing out an inaccuracy in his line "Every moment dies a man—every moment one is born." Babbage explained to Tennyson that since the world population was actually increasing and not, as he indicated, remaining constant, the line should be rewritten to read "Every moment dies a man—every moment one and one-sixteenth is born."

In 1822, Babbage described in a paper and built a model of what he called "a difference engine," which could be used to calculate mathematical tables. The Royal Society of Scientists described his machine as "highly deserving of public encouragement," and a year later the government awarded Babbage £1500 for his project. Babbage originally estimated that the project should take 3 years; however, the design had its complications, and after 10 years of frustrating labor, in which the government grants increased to £17,000, Babbage was still no closer to completion. Finally, the money stopped and Babbage reluctantly decided to let his brainchild go.

In 1833, Babbage developed an idea for a much more practical machine, which he named "the analytical engine." It was to be a more general machine that could be used to solve a variety of problems, depending on instructions supplied by the operator. It would include two units, called a "mill" and a "store," both of which would be made of cogs and wheels. The store, which was equivalent to a modern-day computer memory, could hold up to 100 forty-digit numbers. The mill, which was equivalent to a modern computer's arithmetic and logic unit (ALU), could perform both arithmetic and logic operations on variables or numbers retrieved from the store, and the result could be stored in the store and then acted upon again or printed out. The program of instructions directing these operations would be fed into the analytical engine in the form of punched cards.

The analytical engine was never built. All that remains are the volumes of descriptions and drawings, and a section of the mill and printer built by Babbage's son, who also had to concede defeat. It was, unfortunately for Charles Babbage, a lifetime of frustration to have conceived the basic building blocks of the modern computer a century before the technology existed to build it.

Outline and Objectives

VIGNETTE: BACK TO THE FUTURE

INTRODUCTION

5–1 THE BASICS OF ALGEBRA

Objective 1: Define the following terms.
    a. Algebra
    b. Equation
    c. Formula
    d. Literal numbers

Objective 2: Describe the rules regarding the equality on both sides of the equals sign.

5–1–1 The Equality on Both Sides of the Equals Sign

Objective 3: Demonstrate how to remove a quantity by performing the same arithmetic operation on both sides of an equation or formula.

5–1–2 Treating Both Sides Equally

5–1–3 Directly Proportional and Inversely Proportional

5–2 TRANSPOSITION

5–2–1 Transposing Equations

Objective 4: Explain the steps involved in transposing an equation to determine the unknown.

Objective 5: Demonstrate how to develop an equation from a story problem.

Objective 6: Describe the process of factoring and how to remove parentheses within a formula or expression.

5–2–2 Transposing Formulas

Objective 7: Demonstrate the process of transposing a formula.

Objective 8: Describe how the terms *directly proportional* and *inversely proportional* are used to show the relationship between quantities in a formula.

Objective 9: Explain the proportionality among the three quantities in the Ohm's law formula and describe how to transpose the formula.

Objective 10: Describe the electric power formula and transpose the formula to develop alternative procedures.

5–3 SUBSTITUTION

Objective 11: Demonstrate how to use substitution to develop alternative formulas.

5–3–1 Circle Formula Example

5–3–2 Power Formula Example

5–4 EQUATIONS AND GRAPHING

Objective 12: Perform the following algebra skills:
    a. Create linear equations and express them in standard form
    b. Graph data and linear equations using the slope-intercept form and point-slope form
    c. Describe the properties of exponents
    d. Factor polynomials
    e. Solve quadratic equations using the quadratic formula

5–4–1 Linear Equations

5–4–2 Graphing Data

5–4–3 Graphing Linear Equations

5–4–4 Slope of a Line

5–4–5 Exponents

5–4–6 Factoring Patterns

5–4–7 Solving Quadratic Equations

5–5 RULES OF ALGEBRA—A SUMMARY

Objective 13: Summarize the terms, rules, properties and mathematical operations of algebra.

Introduction

I contemplated calling this chapter "using letters in mathematics" because the word *algebra* seems to make many people back away. If you look up *algebra* in the dictionary, it states that it is "a branch of mathematics in which letters representing numbers are combined according to the rules of mathematics." In this chapter you will discover how easy algebra is to understand and then see how we can put it to some practical use, such as rearranging formulas. As you study this chapter you will find that algebra is quite useful, and because it is used in conjunction with formulas, an understanding is essential for anyone entering a technical field.

5-1 THE BASICS OF ALGEBRA

Algebra
A generalization of arithmetic in which letters representing numbers are combined according to the rules of arithmetic.

Formula
A general fact, rule, or principle expressed usually in mathematical symbols.

As mentioned briefly in the introduction to this chapter, **algebra** by definition is a branch of mathematics in which letters representing numbers are combined according to the rules of mathematics. The purpose of using letters instead of values is to develop a general statement or **formula** that can be used for any values. For example, distance (d) equals velocity (v) multiplied by time (t), or

$$d = v \times t$$

where d = distance in miles
v = velocity or speed in miles per hour
t = time in hours

Using this formula, we can calculate how much distance was traveled if we know the speed or velocity at which we were traveling and the time for which we traveled at that speed. Therefore, if I were to travel at a speed of 20 miles per hour (20 mph) for 2 hours, how far, or how much distance, would I travel? Replacing the letters in the formula with values converts the problem from a formula to an **equation,** as shown:

Equation
A formal statement of the equality or equivalence of mathematical or logical expressions.

$d = v \times t$ ← (Formula)
$d = 20 \text{ mph} \times 2 \text{ h}$ ← (Equation)
$d = 40 \text{ mi}$ ← (Answer)

Literal Number
A number expressed as a letter.

Letters such as d, v, t and a, b, c are called **literal numbers** (letter numbers), and are used, as we have just seen, in general statements showing the relationship between quantities. They are also used in equations to signify an *unknown quantity*, as will be discussed in the following section.

5-1-1 *The Equality on Both Sides of the Equals Sign*

All equations or formulas can basically be divided into two sections that exist on either side of an equals sign, as shown:

Left half   Right half

Fraction bar ⟶ $\boxed{\dfrac{8 \times x}{2}}$ = $\boxed{\dfrac{16}{1}}$ ⟵ Fraction bar

↑
Equals sign

CHAPTER 5 / ALGEBRA, EQUATIONS, AND FORMULAS

Everything in the left half of the equation is equal to everything in the right half of the equation. This means that 8 times *x* (which is an unknown value) divided by 2 equals 16 divided by 1. Generally, it is not necessary to put 1 under the 16 in the right section because any number divided by 1 equals the same number $\left(\frac{16}{1} = 16 \div 1 = 16\right)$; however, it was included in this introduction to show that each section has both a top and a bottom. If the equation is written without the fraction bar or 1 in the denominator position, it appears as follows:

$$\frac{8 \times x}{2} = 16$$

Although the fraction bar is not visible, you should always assume that a number on its own on either side of the equals sign is above the fraction bar, as shown:

$$\frac{8 \times x}{2} = \frac{16}{}$$

Now that we understand the basics of an equation, let us see how we can manipulate it yet keep both sides equal to each other.

5–1–2 *Treating Both Sides Equally*

If you do exactly the same thing to both sides of an equation or formula, the two halves remain exactly equal, or in balance. This means that as long as you add, subtract, multiply, or divide both sides of the equation by the same number, the equality of the equation is preserved. For example, let us try adding, subtracting, multiplying, and dividing both sides of the following equation by 4 and see if both sides of the equation are still equal.

$$\boxed{2 \times 4 = 8} \leftarrow \text{(Original equation)}$$

1. Add 4 to both sides of the equation:

$$2 \times 4 = 8 \quad \leftarrow \text{(Original equation)}$$
$$(2 \times 4) \oplus 4 = 8 \oplus 4 \leftarrow \text{(Add 4 to both sides.)}$$
$$8 + 4 = 8 + 4$$
$$12 = 12$$

Both sides of the equation remain equal.

2. Subtract 4 from both sides of the equation:

$$2 \times 4 = 8 \quad \leftarrow \text{(Original equation)}$$
$$(2 \times 4) \ominus 4 = 8 \ominus 4 \leftarrow \text{(Subtract 4 from both sides.)}$$
$$8 - 4 = 8 - 4$$
$$4 = 4$$

Both sides of the equation remain equal.

3. Multiply both sides of the equation by 4:

$$2 \times 4 = 8 \quad \leftarrow \text{(Original equation)}$$
$$(2 \times 4) \otimes 4 = 8 \otimes 4 \leftarrow \text{(Multiply both sides by 4.)}$$
$$8 \times 4 = 8 \times 4$$
$$32 = 32$$

Both sides of the equation remain equal.

4. Divide both sides of the equation by 4:

$$2 \times 4 = 8 \quad \leftarrow \text{(Original equation)}$$
$$(2 \times 4) \boxed{\div 4} = 8 \boxed{\div 4} \leftarrow \text{(Divide both sides by 4.)}$$
$$\frac{2 \times 4}{4} = \frac{8}{4}$$
$$\frac{8}{4} = \frac{8}{4}$$
$$2 = 2$$

Both sides of the equation remain equal.

As you can see from the four preceding procedures, *if you add, subtract, multiply, or divide both halves of an equation by the same number, the equality of the equation is preserved.*

5–1–3 Ratio and Proportion

Proportional
The relation of one part to another or to the whole with respect to magnitude, quantity, or degree.

The word **proportional** is often used to describe a relationship between two quantities in a formula. For example, consider the following variation on the $d = v \times t$ formula:

$$\text{Time } (t) = \frac{\text{distance } (d)}{\text{velocity } (v)}$$

With this formula we can say that the time (t) it takes for a trip is directly proportional (symbolized by $\propto$) to the distance (d) that needs to be traveled. In symbols, the relationship appears as follows:

$$t \propto d \quad \text{(Time is directly proportional to distance.)}$$

Directly Proportional
The relation of one part to one or more other parts in which a change in one causes a similar change in the other.

The term **directly proportional** can therefore be used to describe the relationship between time and distance because if you increase the distance you have to travel it makes sense that it is going to take more time to travel that distance ($d \uparrow$ causes $t \uparrow$), and similarly, a decrease in distance will cause a corresponding decrease in time ($d \downarrow$ causes $t \downarrow$).

■ **EXAMPLE**

As an example, it will take 5 hours to travel 250 miles at 50 miles per hour:

$$5 \text{ h } (t) = \frac{250 \text{ mi } (d)}{50 \text{ mph } (v)}$$

Let us try doubling the distance and then halving the distance to see what effect it has on time.

■ *Solution:*

If time and distance are directly proportional to each other, they should change in proportion.

$$\text{Original example} \rightarrow \text{Time} = \frac{\text{distance}}{\text{velocity}} = \frac{250 \text{ mi}}{50 \text{ mph}} = 5 \text{ h}$$

$$t = \frac{d}{v} = \frac{\boxed{500 \text{ mi}}}{50 \text{ mph}} = \boxed{10 \text{ h}}$$

Doubling the distance from 250 mi to 500 mi should double the time it takes for the trip, from 5 h to 10 h.

Doubling the distance did double the time ($d \uparrow, t \uparrow$).

Halving the distance from 250 mi to 125 mi should halve the time it takes for the trip, from 5 h to $2\frac{1}{2}$ h.

$$t = \frac{d}{v} = \frac{\boxed{125 \text{ mi}}}{50 \text{ mph}} = \boxed{2.5 \text{ h}}$$

Halving the distance did halve the time $(d\downarrow, t\downarrow)$.

As you can see from this exercise, time is directly proportional to distance ($t \propto d$). Therefore, *formula quantities are always directly proportional to one another when both quantities are above the fraction bars on opposite sides of the equals sign.*

On the other hand, **inversely proportional** means that the two quantities compared are opposite in effect. For example, consider again our time formula:

$$t = \frac{d}{v}$$

Observing the relationship between these quantities again, we can say that the time (t) it takes for a trip is *inversely proportional* (symbolized by $1/\propto$) to the speed or velocity (v) we travel. In symbols, the relationship appears as follows:

$$t \frac{1}{\propto} v \quad \text{(Time is inversely proportional to velocity.)}$$

Inversely Proportional
The relation of one part to one or more other parts in which a change in one causes an opposite change in the other.

The term *inversely proportional* can therefore be used to describe the relationship between time and velocity because if you were to increase your speed or velocity, you would obviously decrease the time it would take for the trip ($v\uparrow$ causes $t\downarrow$), and similarly, a decrease in velocity would cause a corresponding increase in time ($v\downarrow$ causes $t\uparrow$).

■ **EXAMPLE**

Using the previous example,

$$5 \text{ h} (t) = \frac{250 \text{ mi} (d)}{50 \text{ mph} (v)}$$

let us try doubling the speed we travel, or velocity, and then halving the velocity to see what effect it has on time. If time and velocity are inversely proportional to each other, a velocity change should have the opposite effect on time.

■ *Solution:*

$$\text{Original example} \rightarrow \text{Time} = \frac{\text{distance}}{\text{velocity}} = \frac{250 \text{ mi}}{50 \text{ mph}} = 5 \text{ h}$$

Doubling the speed from 50 mph to 100 mph should halve the time it takes to complete the trip, from 5 h to $2\frac{1}{2}$ h.

$$\rightarrow t = \frac{d}{v} = \frac{250 \text{ mi}}{\boxed{100 \text{ mph}}} = \boxed{2.5 \text{ h}}$$

Doubling the speed did halve the time $(v\downarrow, t\uparrow)$.

Halving the speed from 50 mph to 25 mph should double the time it takes to complete the trip, from 5 h to 10 h.

$$\rightarrow t = \frac{d}{v} = \frac{250 \text{ mi}}{\boxed{25 \text{ mph}}} = \boxed{10 \text{ h}}$$

Halving the speed did double the time $(v\downarrow, t\uparrow)$.

As you can see from this exercise, time is inversely proportional to velocity $\left(t \propto \frac{1}{v}\right)$. Therefore, *formula quantities are always inversely proportional to one another when one of the quantities is above the fraction bar and the other is below the fraction bar on opposite sides of the equals sign.*

SELF-TEST EVALUATION POINT FOR SECTION 5–1

Now that you have completed this section, you should be able to:

- **Objective 1.** Define the terms algebra, equation, formula, and literal numbers.
- **Objective 2.** Describe the rules regarding the equality on both sides of the equals sign.
- **Objective 3.** Demonstrate how to remove a quantity by performing the same arithmetic operation on both sides of an equation or formula.

Use the following questions to test your understanding of Section 5–1.

1. Is the following equation true?

$$\frac{\frac{56}{14}}{2} = 2 \quad \text{or} \quad \frac{56 \div 14}{2} = 2$$

2. If we were to multiply the left side of the equation in Question 1 by 3 and the right side of the equation by 2, would the two sides be equal to each other?

3. Fill in the missing values.

$$\frac{144}{12} \times \boxed{} = \frac{36}{6} \times 2 \times \boxed{}$$
$$60 = 60$$

4. Determine the result and state whether the equation is equal.

$$\frac{(8-4) + 26}{5} = \frac{81 - 75}{2}$$

5. Is the following equation balanced?

$$3.2 \text{ k}\Omega = 3200 \ \Omega$$

Use the following technical trade questions to test your understanding of practical applications of Section 5–1.

1. **Electrical** The electrical resistance of a wire is directly proportional to its length, as shown in the example illustration. Wire 1 has a small length and therefore small resistance ($L_1 \downarrow, R_1 \downarrow$). Similarly, wire 2 has a large length and therefore large resistance ($L_2 \uparrow, R_2 \uparrow$). Referring to the equation below, you can see this proportional relationship. The length ratio has wire 1 above the fraction bar, and similarly, the resistance ratio also has wire 1 above the fraction bar—indicating the proportional relationship.

$$\frac{\text{Length of wire 1}}{\text{Length of wire 2}} = \frac{\text{Resistance of wire 1}}{\text{Resistance of wire 2}}$$

The ratio of wire 1's length divided by wire 2's length **is equal to** The ratio of wire 1's resistance divided by wire 2's resistance

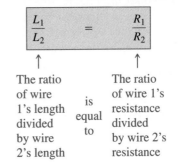

Wire 1
Length = 18 feet
Resistance = 3 ohms

Wire 2
Length = 45 feet
Resistance = 7.5 ohms

a. Insert the values given in the illustration and determine whether both sides of the equation remain equal.
b. If you double the length of wire 1 (36 feet) and so double its resistance (6 ohms), will the equation remain equal?

2. **Automotive** The pulley system shown details the power transfer from an engine to an air conditioning unit. With gears, the speed of a pulley is inversely proportional to its diameter (the larger the diameter, $d \uparrow$, the slower the speed, $s \downarrow$); therefore, the speed of pulley x is inversely proportional to its diameter $\left(s_x = \frac{1}{d_x}\right)$, and the speed of pulley y is inversely proportional to its diameter $\left(s_y = \frac{1}{d_y}\right)$.

The equation below describes this inverse relationship. The speed ratio has pulley x above the fraction bar; conversely, the diameter ratio has pulley x below the fraction bar—indicating the inversely proportional relationship.

$$\frac{\text{Speed of pulley } x}{\text{Speed of pulley } y} = \frac{\text{Diameter of pulley } y}{\text{Diameter of pulley } x}$$

CHAPTER 5 / ALGEBRA, EQUATIONS, AND FORMULAS

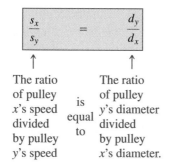

$$\frac{s_x}{s_y} = \frac{d_y}{d_x}$$

↑   ↑

The ratio of pulley x's speed divided by pulley y's speed   *is equal to*   The ratio of pulley y's diameter divided by pulley x's diameter.

a. Insert the following values in the equation and determine whether both sides of the equation remain equal.

Speed of pulley x = 120 revolutions per minute (rpm)
Speed of pulley y = 480 rpm
Diameter of pulley y = 4 inches
Diameter of pulley x = 16 inches

b. If the diameter of pulley x was halved, its speed would be doubled. If these new values were inserted into the equation, would it remain equal?

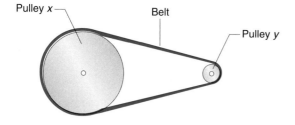

5–2   TRANSPOSITION

It is important to know how to *transpose*, or rearrange, equations and formulas so that you can determine the unknown quantity. This process of rearranging, called **transposition**, is discussed in this section.

> **Transposition**
> The transfer of any term of an equation from one side over to the other side with a corresponding change of the sign.

5–2–1   *Transposing Equations*

As an example, let us return to the original problem introduced at the beginning of this chapter and try to determine the value of the unknown quantity x.

$$\frac{8 \times x}{2} = 16$$

To transpose the equation we must follow two steps:

Step 1.   Move the unknown quantity so that it is above the fraction bar on one side of the equals sign.

Step 2.   Isolate the unknown quantity so that it stands by itself on one side of the equals sign.

Looking at the first step, let us see if our unknown quantity is above the fraction bar on either side of the equals sign.

The unknown quantity x is above the fraction bar. $\dfrac{8 \times \widehat{x}}{2} = 16$

Since step 1 is done, we can move on to step 2. Looking at the equation, you can see that x does not stand by itself on one side of the equals sign. To satisfy this step, we must somehow move the 8 above the fraction bar, and the 2 below the fraction bar away from the left side of the equation, so that x is on its own.

Let us begin by removing the 2. To remove a letter or number from one side of a formula or equation, simply remember this rule: *To move a quantity, simply apply to both sides the arithmetic opposite of that quantity*. Multiplication is the opposite of division, so to remove a "divide by 2" (÷ 2), simply multiply both sides by 2 (× 2), as follows:

$$\frac{8 \times x}{2} = 16 \quad \leftarrow \text{(Original equation)}$$

$$\frac{8 \times x}{2} \times 2 = 16 \times 2 \quad \leftarrow \text{(Multiply both sides by 2.)}$$

$$\frac{8 \times x}{2} \times \frac{2}{1} = 16 \times 2 \quad \leftarrow \text{(Because } \tfrac{2}{2} = 1\text{, the two 2s on the left side of the equation cancel.)}$$

$$(8 \times x) \times 1 = 16 \times 2$$

$$8 \times x = 16 \times 2 \quad \text{(Because anything multiplied by 1 equals the same number, the 1 on the left side of the equation can be removed: } 8x \times 1 = 8x.\text{)}$$

Looking at the result so far, you can see that by multiplying both sides by 2, we effectively moved the 2 that was under the fraction bar on the left side to the right side of the equation above the fraction bar.

$$\frac{8 \times x}{2} = 16 \quad \leftarrow \text{(Original equation)}$$

$$8 \times x = 16 \times 2 \quad \leftarrow \text{(Result after both sides of the equation were multiplied by 2)}$$

However, we still have not completed step 2, which was to isolate *x* on one side of the equals sign. To achieve this we need to remove the 8 from the left side of the equation. Once again we will do the opposite: Because the opposite of multiply is divide, to remove a "multiply by 8" we must divide both sides by 8 (÷ 8), as follows:

$$\frac{8 \times x}{2} = 16 \quad \leftarrow \text{(Original equation)}$$

$$8 \times x = 16 \times 2 \quad \leftarrow \text{(Equation after both sides were multiplied by 2)}$$

$$\frac{8 \times x}{8} = \frac{16 \times 2}{8} \quad \leftarrow \text{(Divide both sides by 8.)}$$

$$\frac{8 \times x}{8} = \frac{16 \times 2}{8} \quad \text{(Because } \tfrac{8}{8} = 1\text{, the two 8s on the left side of the equation cancel.)}$$

$$1 \times x = \frac{16 \times 2}{8} \quad \text{(Anything multiplied by 1 equals the same number, so the 1 on the left side of the equation can be removed: } 1 \times x = x.\text{)}$$

$$x = \frac{16 \times 2}{8}$$

Now that we have completed step 2, which was to isolate the unknown quantity so that it stands by itself on one side of the equation, we can calculate the value of the unknown *x* by performing the arithmetic operations indicated on the right side of the equation.

$$x = \frac{16 \times 2}{8} \quad (16 \times 2 = 32)$$

$$x = \frac{32}{8} \quad (32 \div 8 = 4)$$

$$x = 4$$

To double-check this answer, let us insert this value into the original equation to see if it works.

$$\frac{8 \times x}{2} = 16 \quad \text{(Replace } x \text{ with 4, or substitute 4 for } x.)$$

$$\frac{8 \times 4}{2} = 16 \quad (8 \times 4 = 32)$$

$$\frac{32}{2} = 16 \quad (32 \div 2 = 16)$$

$$16 = 16 \quad \text{(Answer checks out because } 16 = 16.)$$

1. Example with an Unknown Above the Fraction Bar

■ EXAMPLE

Determine the value of the unknown a in the following equation.

$$(2 \times a) + 5 = 23$$

■ *Solution:*

Step 1: Is the unknown quantity above the fraction bar? Yes.

Step 2: Is the unknown quantity isolated on one side of the equals sign? No.

Use the following steps to isolate the unknown, a.

	Steps:
$(2 \times a) + 5 = 23$	To remove the "+ 5," do the opposite: subtract 5 from both sides.
$(2 \times a) + \cancel{5} - \cancel{5} = 23 - 5$	$5 - 5 = 0$; any number plus 0 equals the same number: $(2 \times a) + 0 = 2 \times a$
$2 \times a = 23 - 5$	To remove the "$2 \times$," do the opposite: divide both sides by 2.
$\dfrac{\cancel{2} \times a}{\cancel{2}} = \dfrac{23 - 5}{2}$	$2 \div 2 = 1$; any number multiplied by 1 equals the same number: $1 \times a = a$
$1 \times a = \dfrac{23 - 5}{2}$	
$a = \dfrac{23 - 5}{2}$	Perform the arithmetic operations indicated to determine the value of a. $23 - 5 = 18 \quad 18 \div 2 = 9$
$a = \dfrac{18}{2}$	
$a = 9$	

Double-check your answer by inserting it into the original equation.

$(2 \times a) + 5 = 23$ (Replace *a* with 9.)
$(2 \times 9) + 5 = 23$
$18 + 5 = 23$
$23 = 23$ (Answer checks out because $23 = 23$.)

2. Example with an Unknown Below the Fraction Bar

■ **EXAMPLE**

Determine the value of the unknown *x* in the following equation:

$$\frac{72}{x} = 12$$

■ *Solution:*

Step 1. Is the unknown quantity above the fraction bar? No.

Use the following steps to move *x* above the fraction bar.

$\frac{72}{x} = 12$ *Steps:*
To move the *x* do the opposite:
the opposite of "divide by *x*" is multiply by *x*.

$\frac{72}{\cancel{x}} \cancel{\times x} = 12 \times x$ Any number divided by itself = 1,
$72 \times 1 = 12 \times x$ so $x \div x = 1$ and therefore the two *x*'s
$72 = 12 \times x$ on the left side of the equation cancel.
     $72 \times 1 = 72$

Now that the unknown quantity *x* is above the fraction bar, we can proceed to step 2.

Step 2. Isolate the unknown quantity on one side of the equals sign.

$72 = 12 \times x$ To remove the "12 ×," simply do the opposite: divide both sides by 12.

$\frac{72}{12} = \frac{\cancel{12} \times x}{\cancel{12}}$ $12 \div 12 = 1, 1 \times x = x$; therefore, the 12s cancel.

$\frac{72}{12} = 1 \times x$

$\frac{72}{12} = x$ $72 \div 12 = 6, x = 6$

$6 = x$ or $x = 6$

To double-check your answer, replace *x* with 6 in the original equation.

$\frac{72}{x} = 12$ (Original equation)

$\frac{72}{6} = 12$

$12 = 12$

3. Example with an Unknown on Both Sides of the Equals Sign

In this section we will see how to deal with equations containing an unknown on both sides of the equals sign. In all equations like this, the same letter or literal number will have the same value no matter where it appears.

■ EXAMPLE

Determine the value of the unknown y in the following equation:

$$(6 \times y) + 7 = y + 27$$

■ Solution:

Step 1: Is the unknown above the fraction bar? Yes.

Step 2: Is the unknown isolated on one side of the equals sign? No.

To fulfill step 2 we need to isolate the unknown on one side of the equals sign. Since y exists on both sides, we must remove one. Studying the left side of the equation you can see that it has six y's $(6 \times y)$, whereas the right side of the equation has only one $y(y)$. The first step could therefore be to subtract $y(-y)$ from both sides.

$(6 \times y) + 7 = y + 27$ ← (Original equation)

Steps:

$(6 \times y) + 7\;\boxed{-y} = y + 27\;\boxed{-y}$ — Subtract y from both sides:
$(5 \times y) + 7 = 0 + 27$ — $(6 \times y) - y = 5 \times y,$
$(5 \times y) + 7 = 27$ — $y - y = 0, 0 + 27 = 27$
$(5 \times y) + 7\;\boxed{-7} = 27\;\boxed{-7}$ — To remove $+7$, subtract 7 from both sides.
$\qquad\qquad\qquad\qquad\qquad\quad 7 - 7 = 0, \quad 27 - 7 = 20$
$5 \times y = 20$ — The opposite of multiply is divide. To remove "$5 \times$," divide both sides by 5.
$\dfrac{\cancel{5} \times y}{\cancel{5}} = \dfrac{20}{5}$ — $5 \div 5 = 1$
$1 \times y = \dfrac{20}{5}$ — $1 \times y = y$
$y = \dfrac{20}{5}$ — $20 \div 5 = 4$; therefore, $y = 4$.
$y = 4$

To double-check your answer, replace y with 4 in the original equation.

$(6 \times y) + 7 = y + 27$ ← (Original equation)
$(6 \times 4) + 7 = 4 + 27$ ← (Replace y with 4.)
$24 + 7 = 4 + 27$
$31 = 31$

4. Story Problems

Story problems are probably the best practice because they are connected to our everyday life and help us to see more clearly how we can apply our understanding of mathematics. I have found the following steps helpful with story problems:

a. Determine what is unknown and assign it a literal number such as *x*.
b. Determine how many other elements are involved and how they relate to the unknown.
c. Construct an equation using the unknown and the other elements with their associated arithmetic operations.
d. Calculate the unknown using transposition if necessary.

■ TECHNICAL TRADE APPLICATION: PERSONAL FINANCE

If you multiply the number of dollars in your wallet by 6 and then subtract $23, you will still end up with one fourth of $100. How many dollars do you have in your wallet?

■ *Solution:*

To start with step a, we must determine the unknown. The last sentence, "How many dollars do you have in your wallet?" states quite clearly that this is our unknown quantity. We will let *x* represent the number of dollars in our wallet.

Step b says to determine how many other elements are involved and how they relate. If we read again from the beginning, the example says:

"If you multiply the number of dollars in your wallet by 6 . . .	This part of the story indicates that our unknown *x* should be multiplied by 6; $x \times 6$.
. . . and then subtract $23, . . .	This part states that after our unknown has been multiplied by 6, it should have $23 subtracted; $(x \times 6) - 23$.
. . . you will still end up with . . .	This statement precedes a result and therefore an equals sign should follow our equation at this point; $(x \times 6) - 23 =$.
. . . one-fourth of $100."	This is the result we end up with: $\frac{1}{4}$ of 100 or $\frac{1}{4} \times 100$.

Now, to complete step c, we must construct an equation using the unknown and the other elements with their associated arithmetic operations.

$$(x \times 6) - 23 = \frac{1}{4} \times 100$$

Now that we have an equation, we can move on to step d. Since *x* is not isolated on one side of the equals sign, we will have to transpose or rearrange the equation.

Step 1. Is the unknown above the fraction bar? Yes.

Step 2. Is the unknown isolated on one side of the equals sign? No.

$(x \times 6) - 23 = \frac{1}{4} \times 100$ *Steps:*
Rather than multiply both sides by 4, let us reduce the fraction $\frac{1}{4}$ of 100. 100 is above the fraction bar; therefore, it can be written

$$(x \times 6) - 23 = \frac{1 \times 100}{4} \qquad 1 \times 100 = 100$$

$$(x \times 6) - 23 = \frac{100}{4} \qquad 100 \div 4 = 25; \text{ the fraction is now eliminated.}$$

$$(x \times 6) - 23 = 25$$

$(x \times 6) - 23 \;\boxed{+\,23} = 25 \;\boxed{+\,23}$ Add 23 to both sides to remove -23.

$$-23 + 23 = 0,$$
$$x \times 6 = 48 \qquad (x \times 6) - 0 = x \times 6,$$
$$25 + 23 = 48$$

$$\frac{x \times 6}{\boxed{6}} = \frac{48}{\boxed{6}} \qquad \text{Divide both sides by 6 to remove "multiply by 6."}$$

$$x = \frac{48}{6} \qquad 6 \div 6 = 1, 1 \times x = x$$

$$x = 8$$

You therefore started off with $8 in your pocket. To double-check this answer, replace x with 8 in the original equation.

$$(x \times 6) - 23 = \frac{1}{4} \times 100 \qquad \left(\frac{1}{4} \text{ of } 100 = 25\right)$$
$$(8 \times 6) - 23 = 25 \qquad (8 \times 6 = 48)$$
$$48 - 23 = 25 \qquad (48 - 23 = 25)$$
$$25 = 25$$

■ TECHNICAL TRADE APPLICATION: RETAIL

Store owner A has four times as many nails as store owner B, who has 12,000. Combined, A and B have 60,000 nails. How many nails does store owner A have?

■ *Solution:*

A is the unknown, and as stated in the question, A is four times larger than B.

$$A = 4 \times B$$

If B has 12,000 nails, we can substitute 12,000 for B.

$$A = 4 \times 12,000$$

To calculate how many nails store owner A has, all we have to do is a simple multiplication.

$$A = 4 \times 12,000$$
$$A = 48,000$$

To prove this answer is true, we can return to the part of the question that states "combined, both A and B have 60,000 nails."

$$A + B = 60,000$$

We know A has 48,000 nails and B has 12,000 nails, so we can replace A and B with their equivalent values as follows:

$$A + B = 60,000$$
$$48,000 + 12,000 = 60,000$$
$$60,000 = 60,000$$

5. Factoring (Unmultiplying) and Distributing (Multiplying Through)

A factor is any number multiplied by another number that contributes to the result. For example, consider 3×8. In this multiplication example, 3 is a factor and 8 is a factor, and both will contribute to a result of 24.

Now let us reverse the process. What factors contribute to a result of 12? In this case we must determine which small numbers, when multiplied together, will produce 12. The answer could be

$$2 \times 6 = 12 \quad \text{or} \quad 3 \times 4 = 12$$

Are these, however, the smallest numbers that, when multiplied together, will produce 12? The answer is no because the 6 (in 2×6) can be broken into 2×3, and the 4 (in 3×4) can be broken into 2×2. Therefore, 12 has the factors $2 \times 2 \times 3$.

Now consider the equation $12 + 9$. What smaller number is a factor of both these numbers? The answer is 3 because you can get results of 12 and 9 by multiplying some other number by 3. Therefore, both 12 and 9 contain the factor 3. If we now remove the common factor 3 from both the numbers 12 and 9, we end up with the following equation:

If no symbol appears, it is assumed that it is a multiplication.

$12 + 9 \leftarrow 12 \div 3 = ④, 9 \div 3 = ③$

$④ + ③ \leftarrow$ We cannot let just $4 + 3$ represent $12 + 9$ because the factor 3 is not included.

$3 \times (4 + 3) \quad \text{or} \quad 3(4 + 3) \leftarrow$ This expression will be correct because $3 \times (4 + 3) = 3 \times 7 = 21$ and $12 + 9 = 21$.

Parentheses are used to group the addition $4 + 3$. The factor 3 outside the parentheses applies to everything inside the parentheses.

Factoring is therefore a reducing process that extracts the common factor from two or more larger numbers. As another example, remove the factor from the following equation:

Steps:

$27x - 9x$ 9 is a factor common to both 27 and 9.

$27x \div = 9 = 3x, 9x \div 9 = 1x$

$= 9(3x - 1x)$ Place 9 outside the parentheses and the results of the division inside the parentheses.

We can check if $9(3x - 1x) = 27x - 9x$ simply by performing the arithmetic operations indicated:

$27x - 9x = ⑱x$

$9(3x - 1x) \leftarrow -1x = x$
$= 9(3x - x) \leftarrow 3x - x = 2x$
$= 9 \times 2x$
$= ⑱x$

Results are the same.

CALCULATOR KEYS

Name: Factor function

Function: Some calculators have a factor function that returns an expression factored with respect to all its variables.

factor (expression)

Examples: Extract the common factor from $(64x - 8y)$.
Press keys: Factor $(64x - 8y)$
Answer: $8(8x - y)$
Press keys: Factor $(16a - 4)$
Answer: $4(4a - 1)$

To reverse the factoring process and convert $9(3x - 1x)$ back to the original equation $27x - 9x$, simply reverse the arithmetic steps. Because we began by dividing both numbers by the common factor, simply do the opposite and multiply both values within the parentheses by the factor 9, as follows:

$$9(3x - 1x) = 27x - 9x$$

Steps:
$9 \times 3x = 27x,$
$9 \times 1x = 9x$

To make any equation with parentheses easier to solve, always begin by removing the parentheses before doing any other operation.

This process of removing parentheses, or multiplying through, is called *distributing*, and **the distributive law of multiplication** is as follows.

$$a \times (b + c) = (a \times b) + (a \times c)$$

Distributive Law of Multiplication: The same result is obtained regardless of whether you multiply a whole or multiply each part of the whole.

CALCULATOR KEYS

Name: Expand function

Function: Some calculators have a reverse factoring function that removes the parentheses.

Expand (expression)

Examples:

Press keys: Expand $(8(8x - y))$
Answer: $64x - 8y$
Press keys: Expand $(4(7a + 2b - 4c))$
Answer: $28a + 8b - 16c$

■ **EXAMPLE A**

$$5(7x + 3x) \leftarrow 5 \times 7x = 35x, 5 \times 3x = 15x$$
$$= 35x + 15x$$

■ **EXAMPLE B**

$$6(4a - 3) \quad 6 \times 4a = 24a, \quad 6 \times 3 = 18$$
$$= 24a - 18$$

To summarize, we now know about two new processes called *factoring* and *distributing*. To show how these two are the reverse of each other, let us do a few examples.

■ **EXAMPLE A**

Factoring:

$$12 + 18 \quad \text{Common factor is } 6.$$
$$= 6(2 + 3) \quad 12 \div 6 = 2, 18 \div 6 = 3$$

Distributing:

$$6(2 + 3) \quad 6 \times 2 = 12, 6 \times 3 = 18$$
$$= 12 + 18$$

■ **EXAMPLE B**

Factoring:

$$22y - 11 \quad \text{Common factor is } 11.$$
$$= 11(2y - 1) \quad 22y \div 11 = 2y, 11 \div 11 = 1$$

$$11(2y - 1) \quad \textit{Distributing:}$$
$$= 22y - 11 \quad 11 \times 2y = 22y, 11 \times 1 = 11$$

Now that we know how to perform both of these operations, let us look at a practical application of them.

■ **TECHNICAL TRADE APPLICATION: AUTOMOTIVE**

A motorbike uses a certain number of gallons of gasoline a week, a car uses twice that amount, and a truck uses four times as much as the motorbike and car combined. If all three use 100 gallons of gasoline a week, how much does each vehicle consume per week?

■ *Solution:*

Reading the problem carefully, you can see that each vehicle's consumption is related to the amount consumed by the motorbike. Therefore, if we can find the amount of gasoline consumed by the motorbike, we can calculate the amount consumed by the car and truck. We will represent the unknown quantity of gasoline that the bike consumes by x.

x ← Bike consumption

$2x$ ← Car consumption is twice that of the bike $(2 \times x)$.

$4(x + 2x)$ ← Truck consumption is four times that of the motorbike and car combined $[4 \times (x + 2x)]$. All three use 100 gallons per week.

$$x + 2x + 4(x + 2x) = 100 \text{ gal per week}$$

CHAPTER 5 / ALGEBRA, EQUATIONS, AND FORMULAS

Now that we have an equation, the next step is to use transportation to determine the unknown value x.

Steps:

$x + 2x + 4(x + 2x) = 100$ Distribute:
$4 \times x = 4x, 4 \times 2x = 8x$

$x + 2x + 4x + 8x = 100$ Combine all the unknowns:
$x + 2x + 4x + 8x = 15x$

$15x = 100$ Divide both sides by 15 to remove the "multiply by 15."

$\dfrac{\cancel{15}x}{\cancel{15}} = \dfrac{100}{15}$ $15 \div 15 = 1, 1 \times x = x$

$x = \dfrac{100}{15}$ $100 \div 15 = 6.67$

$x = 6.67$

Therefore,

$$\text{Motorbike consumption} = x = 6.67 \text{ gal per week}$$

The car consumes twice that amount:

$$\text{Car consumption} = 2 \times x = 2 \times 6.67 = 13.34 \text{ gal per week}$$

The truck consumes four times as much as both combined:

$$\text{Truck consumption} = 4(x + 2x) = 4(6.67 + 2 \times 6.67)$$
$$= 4(6.67 + 13.34) \quad 4 \times 6.67 = 26.68$$
$$\quad\quad\quad\quad\quad\quad 4 \times 13.34 = 53.36$$
$$= 26.68 + 53.36$$
$$= 80.0 \text{ gal per week}$$

To double-check our answer, let us add all the individual amounts to see if they equal the total:

$$\text{Motorbike} + \text{car} + \text{truck} = 100 \text{ gal per week}$$
$$6.67 + 13.34 + 80.0 = 100 \text{ gal per week}$$

Referring to the preceding example, you may have noticed that *you used factoring to develop the equation:*

$$\text{Truck consumption is four times that of bike and car} = 4(x + 2x)$$

and *you distributed to transpose the equation*:

$$4(x + 2x) = 4x + 8x$$

6. Square and Square Root

How do we remove the square or square root on one side of an equation? The answer once again is: Perform the opposite function. For example, if we square a number and then find its square root, we end up with the number with which we started:

$$\text{④ squared} = 4^2 = 4 \times 4 = 16$$
$$\text{Square root of } 16 = \sqrt{16} = ④$$

SECTION 5–2 / TRANSPOSITION

■ EXAMPLE

Determine the value of the unknown I in the following equation:

$$I^2 = 81$$

■ Solution:

To remove the square from the unknown quantity I, we simply take the square root of both sides.

$$I^2 = 81$$
$$\sqrt{I^2} = \sqrt{81} \quad \leftarrow \text{(Take square root of both sides.)}$$
$$I = \sqrt{81} \quad \leftarrow \text{(Square and square root cancel.)}$$
$$I = 9 \quad \leftarrow (\sqrt{81} = 9, \text{ since } 9 \times 9 = 81)$$

■ EXAMPLE

Determine the value of the unknown x in the following equation.

$$\sqrt{x} = 2 + 6$$

■ Solution:

$$\sqrt{x} = 2 + 6 \quad \leftarrow \text{(To remove the square root, square both sides.)}$$
$$\sqrt{x^2} = (2 + 6)^2 \quad \leftarrow \text{(Square and square root cancel.)}$$
$$x = (2 + 6)^2$$
$$x = 8^2 \quad (8 \times 8 = 64)$$
$$x = 64$$

5–2–2 Transposing Formulas

As mentioned previously, *a formula is a general statement using letters and sometimes numbers that enables us to calculate the value of an unknown quantity*. Some of the simplest relationships are formulas. For example, to calculate the distance traveled, you can use the following formula, which was discussed earlier.

$$d = v \times t$$

where d = distance in miles
v = velocity in miles per hour
t = time in hours

Using this formula we can calculate distance if we know the speed and time. If we need only to calculate distance, this formula is fine. But what if we want to know what speed to go ($v = ?$) to travel a certain distance in a certain amount of time, or what if we want to calculate how long it will take for a trip ($t = ?$) of a certain distance at a certain speed? The answer is: Transpose or rearrange the formula so that we can solve for any of the quantities in the formula.

■ EXAMPLE

As an example, calculate how long it will take to travel 250 miles when traveling at a speed of 50 miles per hour (mph). To solve this problem, let us place the values in their appropriate positions in the formula.

$$d = v \times t$$
$$250 \text{ mi} = 50 \text{ mph} \times t$$

The next step is to transpose just as we did before with equations to determine the value of the unknown quantity t.

$$250 = 50 \times t \quad \text{To isolate } t, \text{ divide both sides by 50.}$$
$$\frac{250}{50} = \frac{\cancel{50} \times t}{\cancel{50}} \quad 50 \div 50 = 1, 1 \times t = t$$
$$\frac{250}{50} = t \quad 250 \div 50 = 5$$
$$t = 5$$

It will therefore take 5 hours to travel 250 miles at 50 miles per hour ($250 \text{ mi} = 50 \text{ mph} \times 5 \text{ h}$).

All we have to do, therefore, is transpose the original formula so that we can obtain formulas for calculating any of the three quantities (distance, speed, or time) if two values are known. This time, however, we will transpose the formula instead of the inserted values.

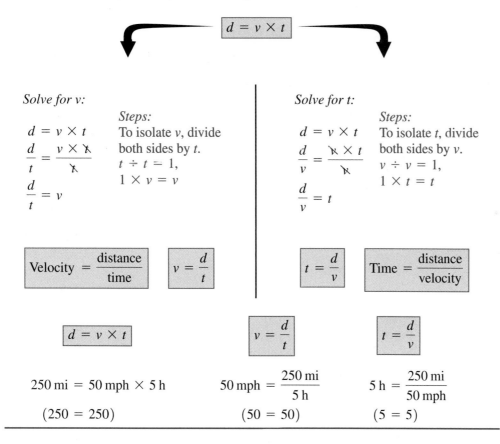

2. Ohm's Law Formula Example

In this section we will examine a law that was first discovered by Georg Ohm in 1826. It states that *the **electric current** in a circuit is directly proportional to the **voltage** source applied and inversely proportional to the **resistance** in the circuit*. Stated as a formula, it appears as follows:

Ohm's law formula $\boxed{\text{Current} = \dfrac{\text{voltage}}{\text{resistance}}}$

Quantity	Symbol	Unit	Symbol
Current	I	Amperes	A
Voltage	V	Volts	V
Resistance	R	Ohms	Ω

Current (electric)
The flow of electronics through a conductor. It is measured in amperes, or amps.

Voltage
Term used to designate electrical power or the force that causes current flow.

Resistance
The opposition to current flow with the dissipation of energy in the form of heat. It is measured in ohms.

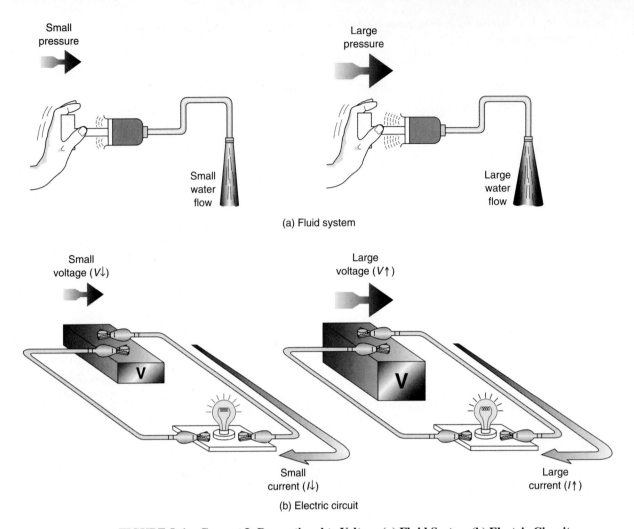

FIGURE 5–1 Current Is Proportional to Voltage (a) Fluid System (b) Electric Circuit.

To explain this formula, let us use the fluid analogy shown in Figure 5–1(a). An object will move only when a force is applied, and as you can see in Figure 5–1(a), the greater the pressure applied to the piston, the greater the amount of water flow through the pipe. Water flow is therefore directly proportional to pressure applied:

$$\text{Water flow} \propto \text{pressure applied}$$

In a similar way, the electric current (I) shown in Figure 5–1(b) is directly dependent on the voltage (V) force applied by the battery. For instance, a smaller voltage $(V \downarrow)$ will cause a corresponding smaller circuit current $(I \downarrow)$, and similarly, a larger voltage $(V \uparrow)$ will result in a larger circuit current $(I \uparrow)$. It can therefore be said that electric current, which is the flow of electrons, is directly proportional to electric voltage, which is the force that moves electrons:

$$\text{Current } (I) \propto \text{voltage } (V)$$

Now, referring to the Ohm's law formula, you can see that current (I) and voltage (V) are directly proportional because these quantities are above the fraction bars on opposite sides of the equals sign.

$$\boxed{\text{Current } (I)} = \frac{\boxed{\text{voltage } (V)}}{\text{resistance } (R)}$$

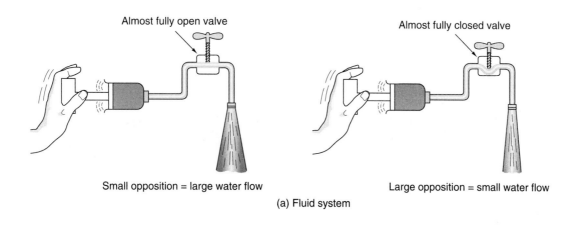
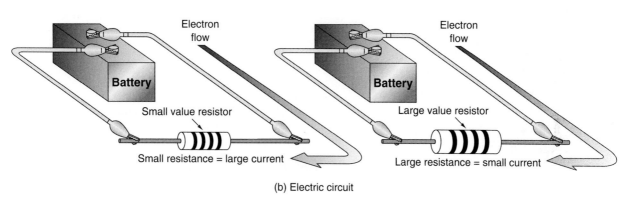

FIGURE 5–2 Current Is Inversely Proportional to Resistance (a) Fluid System (b) Electric Circuit.

The current (I) in an electric circuit is also inversely proportional to the opposition or resistance in the circuit. This can best be explained by the fluid analogy in Figure 5–2(a). If the valve is adjusted to offer a small opposition, a large amount of water will flow, whereas if the valve is adjusted to introduce a large amount of opposition, only a small amount of water will be permitted to flow. Water flow is therefore inversely proportional to the valve's opposition:

$$\text{Water flow} \propto \frac{1}{\text{valve opposition}}$$

In a similar way, the electric current (I) shown in Figure 5–2(b) is inversely dependent on the opposition or resistance (R) offered by the resistor in the circuit. For instance, a small-value resistor will offer only a small opposition or resistance ($R\downarrow$), which will permit a large circuit current ($I\uparrow$), whereas a large-value resistor will offer a large opposition or resistance ($R\uparrow$), which will allow only a small circuit current ($I\downarrow$). It can be said, therefore, that electric current, which is the flow of electrons, is inversely proportional to electric resistance, which is the restriction to current flow:

$$\text{Current } (I) \propto \frac{1}{\text{resistance } (R)}$$

Once again, referring to the Ohm's law formula, you can see that current (I) and resistance (R) are inversely proportional to each other because one quantity is above the fraction bar and the other is below the fraction bar on opposite sides of the equals sign:

$$\boxed{\text{Current } (I)} = \frac{\text{voltage } (V)}{\boxed{\text{resistance } (R)}}$$

■ **EXAMPLE**

Calculate what voltage would cause 2 amperes (A) of electric current in a circuit having an opposition or resistance of 3 ohms (Ω).

■ *Solution:*

To solve this problem, begin by placing the values in their appropriate position within the formula. In this example,

$$\text{Current }(I) = \frac{\text{voltage }(V)}{\text{resistance }(R)}$$

Therefore,

$$2\text{A} = \frac{V?}{3\,\Omega}$$

or

$$2 = \frac{V}{3}$$

As you can see, we will first need to transpose the current formula ($I = V/R$) so that we can find a formula for voltage (V), because V is the unknown in this example.

$$\text{Current }(I) = \frac{\text{voltage }(V)}{\text{resistance }(R)}$$

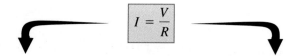

$$I = \frac{V}{R}$$

Solve for V

$$I = \frac{V}{R}$$

$$\frac{R \times I}{R} = \frac{V \times R}{R} \quad \text{(Multiply both sides by } R.\text{)}$$

$$\frac{R \times I}{R} = \frac{V \times \cancel{R}}{\cancel{R}} \quad (R \div R \text{ cancels.})$$

$$\frac{R \times I}{} = \frac{V}{}$$

$$\boxed{V = I \times R}$$

Solve for R

$$I = \frac{V}{R}$$

$$\frac{R \times I}{R} = \frac{V \times R}{R} \quad \text{(Multiply both sides by } R \text{ to bring } R \text{ above the line.)}$$

$$\frac{R \times I}{R} = \frac{V \times \cancel{R}}{\cancel{R}} \quad (R \div R \text{ cancels.})$$

$$\frac{R \times I}{I} = \frac{V}{I} \quad \text{(Divide both sides by } I \text{ to isolate } R.\text{)}$$

$$\frac{R \times \cancel{I}}{\cancel{I}} = \frac{V}{I} \quad (I \div I \text{ cancels.})$$

$$\boxed{R = \frac{V}{I}}$$

Now that we have a formula for voltage ($V = I \times R$), we can complete this problem:

$$V = I \times R \quad \text{(Voltage } = \text{current} \times \text{resistance)}$$
$$= 2\,\text{A} \times 3\,\Omega$$
$$= 6\,\text{V}$$

This equation means that a voltage or electrical pressure of 6 volts would cause 2 amperes of electric current in a circuit having 3 ohms of resistance.

To test if these three formulas are correct, we can now plug into each formula the values of the preceding example, as follows:

$$I = \frac{V}{R} \qquad V = I \times R \qquad R = \frac{V}{I}$$

$$2\,A = \frac{6\,V}{3\,\Omega} \qquad 6\,V = 2\,A \times 3\,\Omega \qquad 3\,\Omega = \frac{6\,V}{2\,A}$$

As you can see from these formulas, the transposition has been successful.

■ TECHNICAL TRADE APPLICATION: ELECTRICAL

Figure 5–3 shows a basic circuit for a home electrical system. If each household light and appliance has 120 volts connected across it, how much current will pass through lamp 1, lamp 2, the hair dryer, and the space heater?

■ *Solution:*

$$\text{Current through lamp 1 } (I_{L1}) = \frac{V_{L1}}{R_{L1}} = \frac{120\,V}{125\,\Omega} = 960\,mA$$

$$\text{Current through lamp 2 } (I_{L2}) = \frac{V_{L2}}{R_{L2}} = \frac{120\,V}{250\,\Omega} = 480\,mA$$

$$\text{Current through hair dryer } (I_{HD}) = \frac{V_{HD}}{R_{HD}} = \frac{120\,V}{40\,\Omega} = 3\,A$$

$$\text{Current through space heater } (I_{SH}) = \frac{V_{SH}}{R_{SH}} = \frac{120\,V}{12\,\Omega} = 10\,A$$

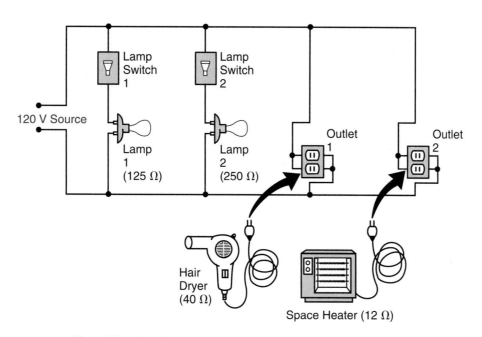

FIGURE 5–3 Home Electrical System.

3. Power Formula Example

Another formula frequently used in electricity and electronics is the power formula. We will also study this formula so that we can practice formula transposition again.

Electric power is the rate at which electric energy is converted into some other form of energy. Power, which is symbolized by P, is measured in *watts* (symbolized by W) in honor of James Watt. We hear watt used frequently in connection with the brightness of a lightbulb: for instance, a 60 watt bulb, a 100 watt bulb, and so on. This designation is a power rating and describes how much electric energy is converted every second. Because a lightbulb generates both heat and light energy, we would describe a 100 watt lightbulb as a device that converts 100 watts of electric energy into heat and light energy every second. Referring to Figure 5–1(b), you can see a lightbulb connected in an electric circuit. The amount of electric power supplied to that lightbulb is dependent on the voltage and the current. It is probably no surprise, therefore, that power is equal to the product of voltage and current. To state the formula:

> Power (Electric)
> The amount of energy converted by a component or circuit in a unit of time, normally seconds. It is measured in watts or joules per second.

$$\boxed{\text{Power} = \text{voltage} \times \text{current} \\ P = V \times I}$$

Quantity	Symbol	Unit	Symbol
Power	P	Watts	W
Voltage	V	Volts	V
Current	I	Amperes	A

This formula states that the amount of power delivered to a device is dependent on the electrical pressure or voltage applied across the device and the electric current flowing through the device.

■ EXAMPLE

What will be the electric current (I) in a circuit if a 60 watt lightbulb is connected across a 120 volt supply voltage?

■ *Solution:*

Let us begin by inserting the known values in their appropriate places in the power formula.

$$\text{Power}(P) = \text{voltage}(V) \times \text{current}(I)$$
$$60\,\text{W} = 120\,\text{V} \times ?\,\text{amperes}$$

As you can see, we first need to transpose the power formula ($P = V \times I$) so that we can find a formula for current (I), because I is the unknown in this example.

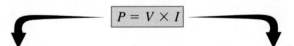

$$P = V \times I$$

Solve for V:

$P = V \times I$

$\dfrac{P}{I} = \dfrac{V \times I}{I}$ (Divide both sides by I.)

$\dfrac{P}{I} = \dfrac{V \times \cancel{I}}{\cancel{I}}$ ($I \div I = 1$, and $V \times 1 = V$.)

$\dfrac{P}{I} = V$

$$\boxed{V = \dfrac{P}{I}}$$

This formula can be used to calculate V when P and I are known.

Solve for I:

$P = V \times I$

$\dfrac{P}{V} = \dfrac{V \times I}{V}$ (Divide both sides by V.)

$\dfrac{P}{V} = \dfrac{\cancel{V} \times I}{\cancel{V}}$ ($V \div V = 1$, and $I \times 1 = I$.)

$\dfrac{P}{V} = I$

$$\boxed{I = \dfrac{P}{V}}$$

This formula can be used to calculate I when P and V are known.

Now that we have a formula for current ($I = P/V$), we can complete this problem:

$$I = \frac{P}{V} \quad \left(\text{Current} = \frac{\text{power}}{\text{voltage}}\right)$$

$$= \frac{60 \text{ W}}{120 \text{ V}}$$

$$= 0.5 \text{ A} \quad \text{or} \quad 500 \times 10^{-3} \text{ A} \leftarrow (0.500.0)$$

$$= 500 \text{ mA}$$

This result means that a current of 500 milliamperes will exist in an electric circuit if a 120 volt battery is connected across a 60 watt light bulb.

CALCULATOR KEYS

Name: Solve function key

Function: Some calculators have a function that solves an expression for a specified variable.

Solve (equation, variable)

Example: Solve for R, when $I = V/R$
Press keys: Solve $(I = V/R, R)$
Answer: $R = V/I$

Solve for t, when $250 = 50 \times t$.
Press keys: Solve $(250 = 50 \times t, t)$
Answer: $t = 5$

Solve for y, when $(6 \times y) + 27 = y + 27$.
Press keys: Solve $((6 \times y) + 7 = y + 27, y)$
Answer: $y = 4$

SELF-TEST EVALUATION POINT FOR SECTION 5–2

Now that you have completed this section, you should be able to:

- **Objective 4.** Explain the steps involved in transposing an equation to determine the unknown.
- **Objective 5.** Demonstrate how to develop an equation from a story problem.
- **Objective 6.** Describe the process of factoring and how to remove parentheses within a formula or expression.
- **Objective 7.** Demonstrate the process of transposing a formula.
- **Objective 8.** Describe how the terms *directly proportional* and *inversely proportional* are used to show the relationship between quantities in a formula.
- **Objective 9.** Explain the proportionality among the three quantities in the Ohm's law formula, and describe how to transpose the formula.
- **Objective 10.** Describe the electric power formula, and transpose the formula to develop alternative procedures.

Use the following questions to test your understanding of Section 5–2.

1. Calculate the result of the following arithmetic operations involving literal numbers.
   a. $x + x =$
   b. $x \times x =$
   c. $7a + 4a =$
   d. $2x - x =$
   e. $\dfrac{x}{x} =$
   f. $x - x =$

2. Transpose the following equations to determine the unknown quantity.
   a. $x + 14 = 30$
   b. $8 \times x = \dfrac{80 - 40}{10} \times 12$
   c. $y - 4 = 8$
   d. $(x \times 3) - 2 = \dfrac{26}{2}$
   e. $x^2 + 5 = 14$
   f. $2(3 + 4x) = (2x + 13)$

3. Transpose the following formulas.
   a. $x + y = z, y = ?$
   b. $Q = C \times V, C = ?$
   c. $x_L = 2 \times \pi \times f \times L, L = ?$
   d. $V = I \times R, R = ?$

SECTION 5–2 / TRANSPOSITION   **143**

4. Determine the unknown in the following equations by using transposition.
   a. $I^2 = 9$
   b. $\sqrt{z} = 8$

Use the following technical trade questions to test your understanding of practical applications of Section 5–2.

1. **Aerospace** The following formula is used to determine the thrust horsepower of a jet engine (H), when the engine is developing T pounds of thrust and is driving the plane at a velocity of V miles per hour.

   $$H = \frac{T \times V}{375}$$

   T = Pounds of thrust
   V = Plane velocity in miles per hour
   H = Thrust horsepower

   a. Calculate H when $T = 12{,}000$ lb and $V = 450$ mph.
   b. Transpose the formula to find V when H and T are known.
   c. Test the formula in (b) with the values given in (a).

2. **Water Management** A water district adds a certain amount of chlorine (in pounds) to a reservoir that has a water flow rate through the reservoir (measured in millions of gallons per day). The formula is as follows:

   $$C_A = 834 \times W \times C_C$$

   C_A = Chlorine added in pounds
   W = Water flow rate in millions of gallons per day
   C_C = Chlorine concentration in parts per million (ppm)

   a. If a reservoir supplies 4.5 million gallons of water per day, and the chlorine concentration should be 24.3 ppm, how many pounds of chlorine should be added?
   b. Transpose the formula to determine chlorine concentration when the water flow rate and the amount of chlorine added is known.
   c. Test the formula in (b) with the values given in (a).

3. **Health Care** The following rule, known as Young's formula, can be employed to determine a child's medicinal dosage.

   $$C_D = \frac{C_A \times A_D}{C_A + 12}$$

   C_D = Child's dose
   C_A = Child's age
   A_D = Adult's dose

   a. Calculate an 8-year-old child's dosage if the adult's dosage is 125 milligrams.
   b. Transpose the formula to determine the adult dosage if the child's age and dosage are known.
   c. Test the formula in (b) with the values given in (a).

4. **Sheet Metal** The following formula determines how long a section of sheet metal should be to accommodate a right-angle inside bend. When lengths 1 and 2 and the thickness of the sheet metal are known:

   $$L_T = L_1 + L_2 + \left(\frac{1}{2} \times T\right)$$

   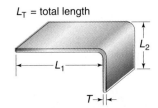

   L_T = total length

   a. Calculate L_T when $L_1 = 10$ cm, $L_2 = 5$ cm, and $I = \frac{1}{2}$ cm.
   b. How much of the total length in (a) is used for the right-angle inside bend?
   c. Solve for L_1 and L_2.

5–3   SUBSTITUTION

Substitution is a mathematical process used to develop alternative formulas by replacing or substituting one mathematical term with an equivalent mathematical term. To explain this process, let us try a couple of examples.

5–3–1   Circle Formula Example

A **circle** is a perfectly round figure in which every point is at an equal distance from the center. Some of the elements of a circle are as follows.

a. The **circumference** (C) of a circle is the distance around the perimeter of the circle.
b. The **radius** (r) of a circle is a straight line drawn from the center of the circle to any point of the circumference.

Substitution
The replacement of one mathematical entity by another of equal value.

Circle
A closed plane curve, every point of which is equidistant from a fixed point within the curve.

Circumference
The perimeter of a circle.

Radius
A line segment extending from the center of a circle or sphere to the circumference.

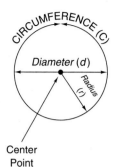
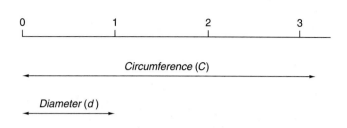

c. The **diameter** (d) of a circle is a straight line drawn through the center of the circle to opposite sides of the circle. The diameter of a circle is therefore equal to twice the circle's radius ($d = 2 \times r$).

Diameter
The length of a straight line passing through the center of an object.

To review, *pi* (symbolized by π) is the name given to the ratio, or comparison, of a circle's circumference with its diameter. This constant will always be equal to approximately 3.14 because the circumference of any circle will always be about 3 times larger than the same circle's diameter. Stated mathematically:

$$\pi = \frac{\text{circumference }(C)}{\text{diameter }(d)} = 3.14$$

■ TECHNICAL TRADE APPLICATION: AUTOMOTIVE

Determine the circumference of a tire that is 26 inches in diameter.

■ *Solution:*

To calculate the wheel's circumference, we will have to transpose the following formula to isolate C.

$$\pi = \frac{C}{d} \qquad \text{Multiply both sides by } d.$$

$$\pi \times d = \frac{C}{d} \times d \qquad d \div d = 1, \text{ and } C \times 1 = C.$$

$$\pi \times d = C$$

The circumference of the bicycle wheel is therefore:

$$C = \pi \times d$$
$$C = 3.14 \times 26 \text{ in.} = 81.64 \text{ inches}$$

In some examples, only the circle's radius will be known. In situations like this, we can substitute the expression $\pi = C/d$ to arrive at an equivalent expression comparing a circle's circumference with its radius.

$$\pi = \frac{C}{d} \qquad \text{Because } d = 2 \times r, \text{ we can substitute } 2 \times r \text{ for } d.$$

$$\pi = \frac{C}{d} \qquad \text{Therefore, this is equivalent to } \pi = \frac{C}{2 \times r}$$

■ TECHNICAL TRADE APPLICATION: LANDSCAPING

How much decorative edging will you need to encircle a garden that measures 6 feet from the center to the edge?

■ *Solution:*

In this example r is known, so to calculate the garden's circumference, we will have to transpose the following formula to isolate C:

$$\pi = \frac{C}{2 \times r} \quad \text{Multiply both sides by } 2 \times r.$$

$$\pi \times (2 \times r) = \frac{C}{2 \times r} \times (2 \times r) \quad (2 \times r) \div (2 \times r) = 1, \text{ and } C \times 1 = C.$$

$$\pi \times (2 \times r) = C$$

The edging needed for the circumference of the garden will therefore be:

$$C = \pi \times (2 \times r)$$
$$C = 3.14 \times (2 \times 6 \text{ ft}) = 37.68 \text{ feet}$$

To continue our example of the circle, how could we obtain a formula for calculating the area within a circle? The answer is best explained with the diagram shown in Figure 5–4. In Figure 5–4(a) a circle has been divided into eight triangles, and then these triangles have been separated as shown in Figure 5–4(b). Figure 5–4(c) shows how each of these triangles can be thought of as having an A and a B section that if separated and rearranged will form a square. The area of a square is easy to calculate because it is simply the product of the square's width and height (area = $w \times h$). The area of the triangles shown in Figure 5–4(c), therefore, will be

$$\text{Area} = \frac{1}{2} \times (w \times h)$$

Referring to Figure 5–4(d), you can see that the width of all eight triangles within a circle is equal to the circle's circumference ($w = C$), and the height of the triangles is equal to the circle's radius ($h = r$). The area of the circle, therefore, can be calculated with the following formula:

$$\text{Area of a circle} = \frac{1}{2} \times (C \times r)$$

Using both substitution and transposition, let us now try to reduce this formula with $C = \pi \times (2 \times r)$.

$$\text{Area of a circle} = \frac{1}{2} \times (C \times r) \quad \leftarrow \text{Because } C = \pi \times (2 \times r), \text{ we can replace } C \text{ with } \pi \times (2 \times r).$$

$$= \frac{1}{2} \times [\pi \times (2 \times r) \times r] \quad \leftarrow \text{We can now use transposition to reduce.}$$

$$= \frac{1 \times \pi \times 2 \times r \times r}{2} \quad \leftarrow 2 \div 2 = 1$$

$$= 1 \times \pi \times r \times r \quad \leftarrow r \times r = r^2$$

$$= 1 \times \pi \times r^2 \quad \leftarrow 1 \times (\pi \times r^2) = \pi \times r^2$$

$$\text{Area of a circle} = \pi \times r^2$$

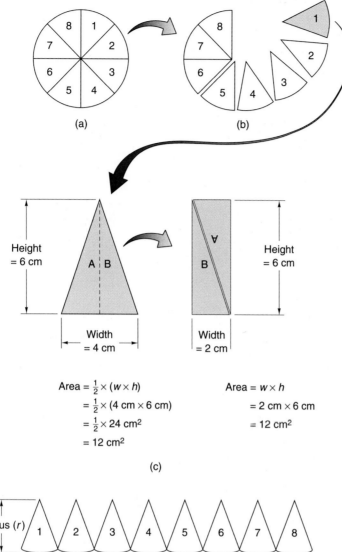

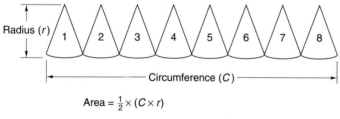

FIGURE 5–4 Calculating the Area of a Circle.

EXAMPLE

Calculate the area of a circle that has a radius of 14 inches.

Solution:

$$\begin{aligned}\text{Area of a circle} &= \pi \times r^2 \\ &= 3.14 \times (14 \text{ in.})^2 \\ &= 3.14 \times 196 \text{ in.}^2 \\ &= 615.4 \text{ square inches}\end{aligned}$$

■ **EXAMPLE**

Calculate the radius of a circle that has an area of 624 square centimeters (cm^2).

■ *Solution:*

To determine the radius of a circle when its area is known, we have to transpose the formula area of a circle $= \pi \times r^2$ to isolate r.

$$\text{Area of a circle} = \pi \times r^2$$
$$A = \pi \times r^2 \quad \leftarrow \text{Divide both sides by } \pi.$$
$$\frac{A}{\pi} = \frac{\pi \times r^2}{\pi} \quad \leftarrow \pi \div \pi = 1, 1 \times r^2 = r^2$$
$$\frac{A}{\pi} = r^2 \quad \leftarrow \text{Take the square root of both sides.}$$
$$\sqrt{\frac{A}{\pi}} = \sqrt{r^2} \quad \leftarrow \sqrt{r^2} = r$$
$$\sqrt{\frac{A}{\pi}} = r$$

Now we can calculate the radius of a circle that has an area of 624 cm^2.

$$r = \sqrt{\frac{A}{\pi}} = \sqrt{\frac{624}{3.14}} = \sqrt{198.7} = 14.1 \text{ centimeters}$$

5–3–2 Power Formula Example

In Section 5–2 you were introduced to two electrical formulas: the Ohm's law formula, which states

$$\text{Current }(I) = \frac{\text{voltage }(V)}{\text{resistance }(R)}$$

and the power formula, which states

$$\text{Power }(P) = \text{voltage }(V) \times \text{current }(I)$$

Using transposition we were able to develop the following alternative formulas that allowed us to calculate any quantity if two quantities were known.

OHM'S LAW FORMULA			POWER FORMULA		
FORMULA	CALCULATES	WHEN THESE ARE KNOWN:	FORMULA	CALCULATES	WHEN THESE ARE KNOWN:
$I = \frac{V}{R}$	Current (I) in amperes (A)	Voltage (V) and resistance (R)	$P = V \times I$	Power (P) in watts (W)	Voltage (V) and current (I)
$V = I \times R$	Voltage (V) in volts (V)	Current (I) and resistance (R)	$V = \frac{P}{I}$	Voltage (V) in volts (V)	Power (P) and current (I)
$R = \frac{V}{I}$	Resistance (R) in ohms (Ω)	Voltage (V) and current (I)	$I = \frac{P}{V}$	Current (I) in amperes (A)	Power (P) and voltage (V)

Although it seems from this table that we have covered all possible alternatives, there are a few missing. For example, consider the power formula $P = V \times I$. When trying to calculate power, we may not always have voltage (V) and current (I). We may, for example, have values only for voltage (V) and resistance (R), or current (I) and resistance (R). In situations such as this the mathematical process called *substitution* can be used to develop alternative power formulas by replacing or substituting one mathematical term for an equivalent mathematical term. The following process shows how we can substitute terms in the $P = V \times I$ formula to arrive at alternative power formulas for wattage calculations.

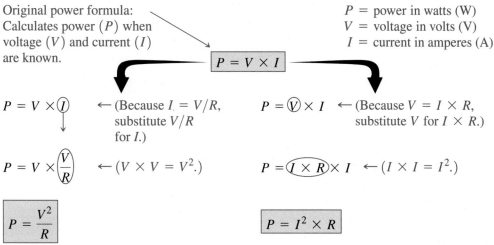

This formula calculates power (P) when voltage (V) and resistance (R) are known.

This formula calculates power (P) when current (I) and resistance (R) are known.

Let us now look at two applications to see how we can use these formulas to calculate unknown quantities.

■ TECHNICAL TRADE APPLICATION: ELECTRICAL

The household heater shown here has a resistance of 7 ohms and is connected to the 120 volt source at the wall outlet. Calculate the amount of power dissipated by the heater.

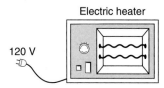

Electric heater

120 V

■ Solution:

In this example, both voltage (120 V) and resistance (7 Ω) are known, and the unknown to be calculated is power (P). We must therefore look at all the power formulas to select one that includes voltage (V) and resistance (R).

$$P = V \times I$$
$$P = \frac{V^2}{R} \leftarrow \text{This power formula is the one to use because it includes } V \text{ and } R.$$

SECTION 5–3 / SUBSTITUTION   **149**

$$P = I^2 \times R$$

$$\text{Power }(P) = \frac{\text{voltage}^2\,(V^2)}{\text{resistance}\,(R)}$$

Replace the literal numbers V and R with values.

$$= \frac{(120\text{ V})^2}{7\,\Omega}$$

$120^2 = 14{,}400$

$$= \frac{14{,}400}{7\,\Omega}$$

$$= 2057\text{ W}$$

$2.\overset{\frown}{057}$

or approximately 2 kW

■ TECHNICAL TRADE APPLICATION: ELECTRICAL

The lightbulb shown here is rated at 6 volts, 60 milliamps.

    a. Using Ohm's law, calculate the resistance of the electric lamp or lightbulb.
    b. Using the power formula, calculate the lamp's wattage at its rated current and resistance.

■ *Solution:*

Part (a) asks us to use Ohm's law to calculate the resistance of the electric lamp.

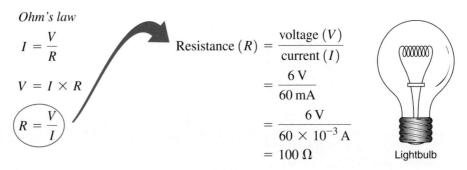

Part (b) directs us to use the power formula (P) to calculate the lamp's wattage at its rated current (I) and resistance (R). I and R are known, so we will use the following formula:

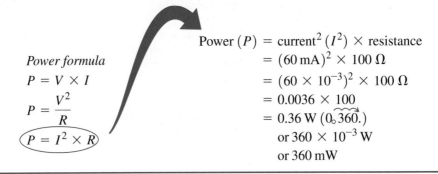

As a review of the Ohm's law and power formula combinations, Figure 5–5 shows all the possible formulas for the four electrical properties.

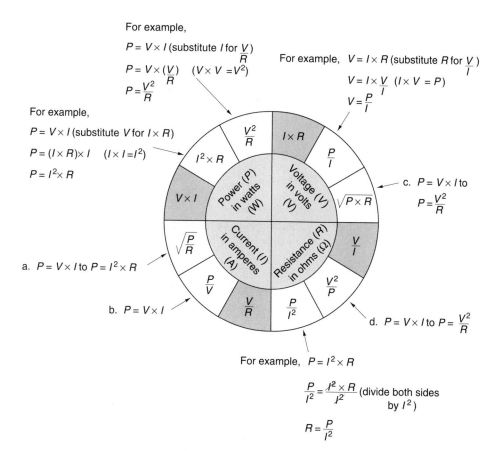

FIGURE 5–5 The Ohm's Law and Power Formula Circle.

SELF-TEST EVALUATION POINT FOR SECTION 5–3

Now that you have completed this section, you should be able to:

■ *Objective 11.* *Demonstrate how to use substitution to develop alternative formulas.*

Use the following questions to test your understanding of Section 5–3.

1. Use substitution to calculate the value of the unknown.

$$x = y \times z \quad \text{and} \quad a = x \times y$$

   Calculate a if $y = 14$ and $z = 5$.

2. If $I = V \times R$ and $P = V \times I$, show the transposition and substitution steps that were followed to develop the formula circle shown in Figure 5–5. (The development follows parts (a) to (d) in the figure.)

3. What percentage of 12 gives 2.5?

$$(x\% \text{ of } 12 = 2.5)$$

4. 60 is 22% of what number?

$$(22\% \text{ of } x = 60)$$

Use the following technical trade questions to test your understanding of practical applications of Section 5–3.

1. **Machining** Pomeroy's formula can be used to approximate the power needed by a metal punch machine:

$$P \approx \frac{T^2 \times d \times N}{3.78}$$

where
P = power required (in horsepower)
T = metal thickness
d = hole diameter
N = number of holes to be punched simultaneously
$\approx$: approximately equal to
d = hole diameter

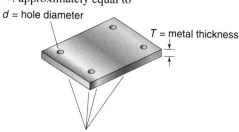

All four will be punched simultaneously ($N = 4$)

a. If the hole's radius is known instead of its diameter, what would be the new formula?
b. Transpose Pomeroy's formula to determine how many holes a certain punch machine can manage (solve for N).
c. Determine P when:

$$I = 3 \text{ cm}$$
$$d = 0.5 \text{ cm}$$
$$N = 4$$

2. **Plumbing** The following formula is used to calculate the wall thickness of the piping shown:

$$W_T = \frac{1}{2}(d_O - d_I)$$

If the inside diameter is $\frac{2}{3}$ of the outside diameter ($d_I = \frac{2}{3} \times d_O$):
a. Write the formula to calculate W_T, when only d_O is known.
b. Calculate W_T when $d_O = 2\frac{1}{4}$ inches.

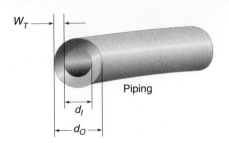

Piping

W_T = Wall thickness
d_I = Inside diameter
d_O = Outside diameter

c. Simplifying the formula in (a), we can obtain an equivalent formula that is:

$$W_T = \frac{d_O}{6}$$

Test if this can be used by again using $2\frac{1}{4}$ inches as in (b).

5–4 EQUATIONS AND GRAPHING

As stated previously, when algebraic terms or expressions are separated by an equals sign, the resulting algebraic statement is called an equation.

5–4–1 *Linear Equations*

Literal Number
A letter used to represent quantities or variables. If a salad has tomatoes, lettuce, and cucumbers, $S = T + L + C$.

Linear Equation
If the literal numbers in an equation are raised to the first power, they are called linear equations. The first power is not normally included, since $b^1 = b$.

Standard Form
A linear equation can appear to be very different from another equation when in fact they are equal. To avoid confusion, a standard (STF) was adopted: $ax + by = c$.

Coefficient
The numerical factors in a term.

Integer
A number without a fractional part.

In the simplest equations, the **literal number** (letters used to represent quantities) are raised to the first power, and are referred to as **linear equations.** For example:

$$d = v \times t$$ ← Each of these literal numbers is raised to the first power:
$$d^1 = v^1 \times t^1$$
This first power is not normally included since $d^1 = d$.

A linear equation can appear to be different, but in fact be exactly the same. For instance, $x + y = 2$ is the same as $x = 2 - y$. To avoid confusion a **standard form** (STF) was adopted so that an alternate correct answer was not mistaken for an incorrect one.

Standard form (STF): $Ax + By = C$

Variable terms are on the left of the equals sign.
Number is always on the right of the equals sign.
Coefficients (A, B, C) must be integers.
A is generally always positive.

The numerical **coefficient** of an algebraic term is the real number part of the term. For example, with the term $6x$, the number 6 is the numerical coefficient. An **integer** is a whole number without a fractional part, including zero. For example, 3, 0, −6, and +12 are all examples of integers, whereas 18.3 and $-6\frac{1}{2}$ are not integers because they have fractional parts.

■ **EXAMPLE:**

Convert the following linear equation into standard form.

$$3x - 4y - 3 = 6x + 5y - 8$$

■ *Solution:*

$$3x - 4y - 3 = 6x + 5y - 8 \quad (-3x)$$
$$-4y - 3 = 3x + 5y - 8 \quad (+4y)$$
$$-3 = 3x + 9y - 8 \quad (+8)$$
$$+5 = 3x + 9y$$
$$\text{STF:} \quad 3x + 9y = 5$$

5–4–2 Graphing Data

In many areas of science and technology, you will often see data shown visually in the form of a graph. To obtain this data, a *practical analysis* is performed on a device, in which the input is physically changed in increments and the output results are recorded in a table.

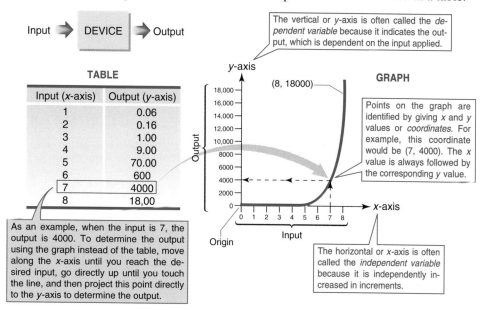

The graph shown would accommodate any data in which the input and output data are positive, but what sort of graph could be used if positive and negative values had to be plotted? The answer is a **Cartesian coordinate system**. With this system, the previous positive vertical x-axis and horizontal y-axis are still used, but are extended beyond the zero point in the opposite direction to accommodate negative values.

Cartesian Coordinate System

Two perpendicular number lines used to place points in a plane.

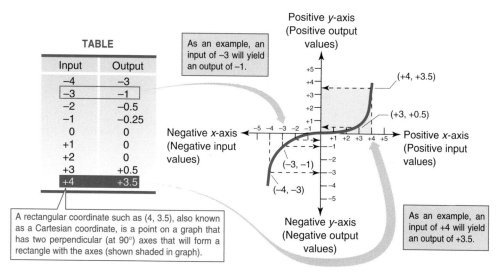

SECTION 5–4 / EQUATIONS AND GRAPHING

5-4-3 Graphing Linear Equations

To combine our understanding of linear equations and graphs, let us step through the process of converting an equation to a graph.

Consider the linear equation:

$$y = 2x + 2$$

This equation is, in fact, the formula you would use to determine the output (y) from a device for any given input (x). For example, if you applied 3 to the input $(x = 3)$, the output would be:

$$y(\text{output}) = 2(3) + 2 \qquad x(\text{input}) = 3$$
$$= 6 + 2$$
$$= 8$$

So how can we graph the response of this device? The answer is to pick some example input values, calculate the outputs for these inputs, and plot the points on a graph.

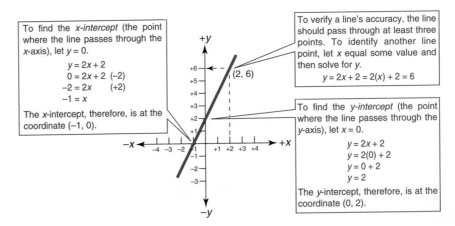

5-4-4 Slope of a Line

The steepness of the line in a graph is an indication of how much or little the output is changing for a given input. For example, in the distance versus time graph that follows, we can see how fast we were going by seeing how steep the line is.

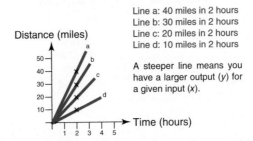

Line a: 40 miles in 2 hours
Line b: 30 miles in 2 hours
Line c: 20 miles in 2 hours
Line d: 10 miles in 2 hours

A steeper line means you have a larger output (y) for a given input (x).

The **slope of a line** is defined as the change in y that results from an increase in x.

$$\text{Slope} = \frac{\Delta y}{\Delta x}$$

Δy = change in y, called *rise*

Δx = change in x, called *run*

Change is generally symbolized by the Greek letter delta, Δ.

Slope of a Line
The ratio of the vertical change to the horizontal change between two points plotted on a line.

The change in input (Δx) value will always be positive, but the change in output (Δy) value can be either positive or negative based on whether y increased or decreased. To explain this point, take a look at the following two examples.

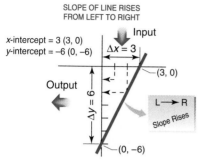

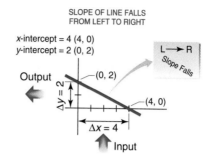

When a slope rises from left to right, the slope is said to be positive, and so y should be a positive value.

When a slope falls from left to right, the slope is said to be negative, and so y should be a negative value.

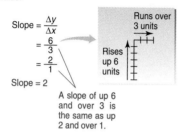

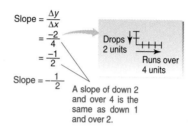

Any linear equation containing two variables can be written in the **slope-intercept form (SIF)**.

$$\text{Slope-Intercept Form (SIF):} \quad y = mx + b$$

$$m = \text{slope}$$
$$b = y\text{-intercept}$$

Slope-Intercept Form
A form used to express the slope of a line in which m = slope and b indicates the y-intercept,
SIF: $y = mx + b$.

In this appropriately named form, m represents the slope ($\Delta y / \Delta x$) and b the y-intercept. The main characteristic of an equation in slope-intercept form is that it is solved for y, which means that y should be by itself on the left side of the equation.

■ **EXAMPLE**

Referring to the graph, determine the equation for the line and give it in slope-intercept form.

■ *Solution:*

Slope rises from left to right and so y is positive.

$$\text{Slope} = \frac{\Delta y}{\Delta x} = \frac{2}{1} = 2$$

$$y\text{-intercept} = +2$$

The equation of the line in slope-intercept form is:

$$(\text{SIF}) \quad y = 2x + 2$$

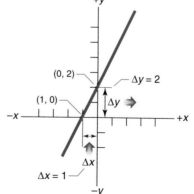

■ **EXAMPLE**

Convert the following equation to standard form (STF) and slope-intercept form (SIF), and then graph the line.

$$3y - 2x = 9$$

SECTION 5–4 / EQUATIONS AND GRAPHING

■ *Solution:*

First, convert the equation to standard form (STF), which is $Ax + By = C$.

$$3y - 2x = 9 \quad (+2x)$$
$$3y = 9 + 2x \quad (-9)$$
$$3y - 9 = 2x \quad (-3y)$$
$$-9 = 2x - 3y$$
$$\text{STF:} \quad 2x - 3y = -9$$

Now, convert to slope-intercept form (SIF), which is: $y = mx + b$.

$$3y - 2x = 9 \quad (+2x)$$
$$3y = 9 + 2x \quad (\div 3)$$
$$y = \frac{2x + 9}{3}$$
$$y = \frac{2}{3}x + \frac{9}{3}$$
$$\text{SIF:} \quad y = \frac{2}{3}x + 3$$

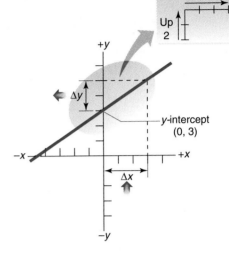

Using the SIF, it is easy to graph the line, as shown. The y-intercept is $+3$, the slope is positive $\left(\frac{2}{3}\right)$, and so the line rises from left to right, up 2 from the y-intercept and over 3.

Point-Slope Form
A form used to express the slope of a line in which x_1, y_1 indicates the point and m indicates slope.
PSF: $y - y_1 = m(x - x_1)$.

Linear equations can also be represented in **point-slope form (PSF),** in which the point given may not be the y-intercept.

Point-Slope Form (PSF): $\quad y - y_1 = m(x - x_1) \quad m = \text{slope}$

Subscripts (y_1 and x_1) are used to denote the point you are given.

■ **EXAMPLE**

The line shown in the graph crosses through point $(-5, 2)$ and has a slope of $-\frac{1}{5}$. What is the equation for this line in point-slope form?

■ *Solution:*

Point $(-5, 2)$ is not the y-intercept, $x_1 = -5$, $y_1 = 2$, and $m = -\frac{1}{5}$. Therefore:

$$y - y_1 = m(x - x_1)$$
$$y - 2 = -\frac{1}{5}(x - (-5))$$
$$y - 2 = -\frac{1}{5}(x + 5)$$
$$\text{PSF:} \quad y - 2 = -\frac{1}{5}(x + 5)$$

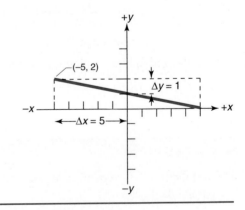

■ **EXAMPLE**

Convert the point-slope form (PSF) equation in the previous example into standard form (STF).

■ *Solution:*

$$\text{PSF:} \quad y - 2 = -\frac{1}{5}(x + 5) \qquad \text{STF:} \quad Ax + Bx = c$$

To convert to standard form (STF), the coefficients must be integers (no fractions).

$$y - 2 = -\frac{1}{5}(x + 5) \qquad (\times 5)$$
$$5y - 10 = -1(x + 5) \qquad \text{(distribute the negative, } -1 \cdot (x + 5))$$
$$5y - 10 = -x - 5 \qquad (+10)$$
$$5y = -x + 5 \qquad (+x)$$
$$x + 5y = 5$$
$$\text{STF:} \quad x + 5y = 5$$

5-4-5 *Exponents*

Many literal numbers are raised to a power, and so it is important to understand how to deal with these exponents if you are to transpose equations and formulas. In this section, we will review some of the rules discussed previously, and examine some new equalities:

1. Multiplication of Exponents

$$a^2 \cdot a^3 = a^{2+3} = a^5$$

Explanation:

$$a^2 \cdot a^3 = (a \cdot a) \cdot (a \cdot a \cdot a)$$
$$= a \cdot a \cdot a \cdot a \cdot a = a^5 \qquad (\text{"}a\text{" as a factor 5 times})$$

2. Division of Exponents

$$\frac{a^3}{a^2} = a^{3-2} = a^1 = a$$

Explanation:

$$\frac{a^3}{a^2} = \frac{a \cdot \cancel{a} \cdot \cancel{a}}{\cancel{a} \cdot \cancel{a}} = a \cdot 1 \cdot 1 = a \qquad (a \div a = 1)$$

3. Negative Exponents

$$a^{-2} = \frac{1}{a^2}$$

Explanation:

$$2^2 = 2 \cdot 2 = 4$$
$$2^{-2} = \frac{1}{2} \cdot \frac{1}{2} = \frac{1}{4} = \frac{1}{2^2}$$

4. Exponents Raised to a Power

$$(a^3)^2 = a^{3\times 2} = a^6$$

Explanation:

$$(a^3)^2 = (a \cdot a \cdot a)^2 = (a \cdot a \cdot a) \cdot (a \cdot a \cdot a) = a \cdot a \cdot a \cdot a \cdot a \cdot a = a^6$$

5. Root Power

$$\sqrt[3]{a^2} = \left(\sqrt[3]{a}\right)^2$$

Example:

$$\sqrt[3]{64^2} = \left(\sqrt[3]{64}\right)^2$$
$$\sqrt[3]{4096} = (4)^2$$
$$16 = 16$$

Whether you square "*a*" and then find the cube root of the result, or find the cube root of "*a*" and then square the result, the final answer is the same.

6. Fractional Power

$$a^{\frac{2}{3}} = \sqrt[3]{a^2}$$

Explanation:

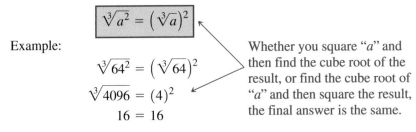

■ **EXAMPLE**

The reciprocal of a number will perform the opposite or inverse action on a value. For instance:

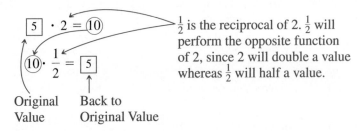

$\frac{1}{2}$ is the reciprocal of 2. $\frac{1}{2}$ will perform the opposite function of 2, since 2 will double a value whereas $\frac{1}{2}$ will half a value.

Original Value   Back to Original Value

Therefore, since $\frac{1}{3}$ is the reciprocal of 3:

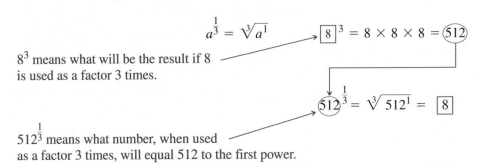

8^3 means what will be the result if 8 is used as a factor 3 times.

$512^{\frac{1}{3}}$ means what number, when used as a factor 3 times, will equal 512 to the first power.

5-4-6 Factoring Patterns

In this section, we will take our understanding of factoring a step further.

1. Basic Patterns

A three term (trinomial) expression can be factored if you can find two numbers whose sum is the coefficient of the second term, and whose product is the third term.

$$x^2 + bx + c = (x + m) \cdot (x + n)$$

■ **EXAMPLE:**

$$x^3 + 13x + 40 = (x + ?) \cdot (x + ?)?$$

■ *Solution:*

In this expression, the coefficient of the second term is 13 and the coefficient of the third term is 40. Using trial and error, test to see if you can find two numbers whose sum $(+)$ is 13, and whose product $(\times)$ is 40.

$$12 + 1 = 13 \,(\text{Yes}), \quad 12 \times 1 = 12 \,(\text{No})$$
$$11 + 2 = 13 \,(\text{Yes}), \quad 11 \times 2 = 22 \,(\text{No})$$
$$10 + 3 = 13 \,(\text{Yes}), \quad 10 \times 3 = 30 \,(\text{No})$$
$$9 + 4 = 13 \,(\text{Yes}), \quad 9 \times 4 = 36 \,(\text{No})$$
$$8 + 5 = 13 \,(\text{Yes}), \quad 8 \times 5 = 40 \,(\text{Yes})$$

The two common factors for 13 and 40, therefore, are 8 and 5. These can be placed in the factoring pattern as follows:

$$x^2 + 13x + 40 = (x + 8) \cdot (x + 5)$$

To test if the factoring pattern in the previous example is correct, let us multiply through to see if we obtain our original equation.

		Steps:
$(x + 8) \cdot (x + 5)$	$x^2 + ? + ? + ?$	$x \cdot x = x^2$
$(x + 8) \cdot (x + 5)$	$x^2 + 5x + ? + ?$	$x \cdot 5 = 5x$
$(x + 8) \cdot (x + 5)$	$x^2 + 5x + 8x + ?$	$8 \cdot x = 8x$
$(x + 8) \cdot (x + 5)$	$x^2 + 5x + 8x + 40$	$8 \cdot 5 = 40$

$$(x + 8) \cdot (x + 5) = x^2 + (5x + 8x) + 40 = x^2 + 13x + 40$$

The order in which 8 and 5 are inserted into the factoring pattern is immaterial as proved here:

$$(x + 8)(x + 5) \qquad \text{or} \qquad (x + 5)(x + 8)$$
$$= x^2 + 5x + 8x + 40 \qquad\qquad = x^2 + 8x + 5x + 40$$
$$= x^2 + 13x + 40 \qquad\qquad = x^2 + 13x + 40$$

In summary, here are the factoring pattern combinations, all of which can be obtained by using the "Factor" function on your calculator, as previously described.

Factoring Patterns (Trinomials)

▢ + ▢ + ▢ = $(x + m)(x + n)$

▢ + ▢ − ▢ = $(x + m)(x − n)$ or $(x − m)(x + n)$

▢ − ▢ + ▢ or ▢ − ▢ − ▢ = $(x − m)(x − n)$

2. Difference of Perfect Squares

A perfect square is a number such as 16 or 25, since it is a result of a number multiplied by itself. For example:

$$4 \times 4 = 16$$
$$5 \times 5 = 25$$
Perfect squares

If you have one perfect square subtracted (difference) from another, you can factor it using the following pattern.

$$x^2 - y^2 = (x - y)(x + y)$$

Example:

$$x^2 - 25$$
$$= x^2 - 5^2 \quad (x^2 \text{ and } 5^2 \text{ are perfect squares})$$
$$= (x - 5)(x + 5)$$

You can factor the difference of perfect squares $[x^2 - 4 = (x - 2)(x + 2)]$, but you cannot factor the sum of perfect squares $(x^2 + 4 = x^2 + 4)$.

3. Sum of Perfect Cubes

$$x^3 + y^3 = (x + y)(x^2 - xy + y^2)$$

Example:

$$x^3 + 125$$
$$= x^3 + 5^3$$
$$= (x + 5)(x^3 - 5x + 5^2)$$

4. Difference of Perfect Cubes

$$x^3 - y^3 = (x - y)(x^2 + xy + y^2)$$

Example:

$$x^3 - 64$$
$$= x^3 - 4^3$$
$$= (x - 4)(x^2 + 4x + 4^2)$$

5–4–7 Solving Quadratic Equations

As stated previously, if the literal numbers in an equation are raised to the first power, they are called linear equations. If the literal numbers are raised to the second power, they are called **second-degree equations** or **quadratic equations.** In standard form (STF) a quadratic equation looks like this:

$$\text{STF:} \quad ax^2 + bx + c = 0 \quad \text{(Quadratic Equation)}$$

a is the coefficient of x^2

b is the coefficient of x

c is a constant

Whenever you solve for the unknown (x) in a quadratic equation, there will always be one of two possible answers. To prove this, let us look at an example.

Quadratic or Second-Degree Equations
If the literal numbers in an equation are raised to the first power, they are called linear equations. If the literal numbers are raised to the second power, they are called second-degree or quadratic equations.

■ **EXAMPLE**

Solve the following equation:
$$3x^2 + 4x = -1$$

■ *Solution:*

The first step is to set the equation so that it is equal to zero:

$$3x^2 + 4x = -1$$
$$(3x + 1) \cdot (x + 1) = 0$$

Since the product of these two groups is equal to zero, we know that one of the groups is equal to zero:

$(3x + 1) \cdot (0) = 0 \qquad (0) \cdot (x + 1) = 0$
$x = -\dfrac{1}{3} \qquad\qquad x = -1$

$(a) \cdot (0) = 0$
$(0) \cdot (b) = 0$

This equation, like all quadratic equations, has two possible solutions $\left(-\dfrac{1}{3} \text{ and } -1\right)$. To see if this is true, insert each into the original equation $(3x^2 + 4x = -1)$.

When your quadratic equation is set to zero $(ax^2 + bx + c = 0)$, you can solve for x by inserting your a, b, and c coefficients into the following **quadratic formula.**

$$\boxed{x = \dfrac{-b \pm \sqrt{b^2 - 4ac}}{2a}}$$

Quadratic Formula
A formula used to determine the unknown x for a quadratic or second-degree equation.

■ **EXAMPLE**

Solve the equation $2x^2 + 12x - 18 = 0$

■ *Solution:*

$$a = 2, b = 12, c = -18$$
$$x = \dfrac{-b \pm \sqrt{b^2 - 4ac}}{2a}$$
$$= \dfrac{-(12) \pm \sqrt{(12)^2 - 4(2)(-18)}}{2(2)}$$
$$= \dfrac{-12 \pm \sqrt{144 - (-144)}}{4}$$
$$= \dfrac{-12 \pm \sqrt{288}}{4}$$
$$= \dfrac{-12 \pm 12\sqrt{2}}{4}$$

$= \dfrac{(-12) + (12\sqrt{2})}{4} \qquad = \dfrac{(-12) - (12\sqrt{2})}{4}$

$= \dfrac{-12}{4} + \dfrac{12\sqrt{2}}{4} \qquad = \dfrac{-12}{4} - \dfrac{12\sqrt{2}}{4}$

$= (-3) + (3\sqrt{2}) \qquad = (-3) - (3\sqrt{2})$

Period
Time taken to complete one complete cycle of a periodic or repeating waveform.

Frequency
Rate or recurrences of a periodic wave normally within a unit of one second, measured in hertz (cycles/second).

The following memory aid may help you to remember the equation formula.

The bad boy $(-b)$	$\rightarrow x = -b$
could not decide yes or no $(\pm)$	$\rightarrow x = -b \pm$
on whether to go to the radical party $(\sqrt{\ })$	$\rightarrow x = -b \pm \sqrt{\ }$
where he could square dance	$\rightarrow x = -b \pm \sqrt{b^2}$
to the bad music of the Four Aces	$\rightarrow x = -b \pm \sqrt{b^2 - 4ac}$
until it was all over at 2 a.m.	$\rightarrow x = \dfrac{-b \pm \sqrt{b^2 - 4ac}}{4}$

SELF-TEST EVALUATION POINT FOR SECTION 5–4

Now that you have completed this section, you should be able to:

■ **Objective 12.** *Perform the following algebra skills:*
   a. *Create linear equations and express them in standard form*
   b. *Graph data and linear equations using the slope-intercept form and point-slope form*
   c. *Describe the properties of exponents*
   d. *Factor polynomials*
   e. *Solve quadratic equations using the quadratic formula*

Use the following questions to test your understanding of Section 5–4.

1. Write the equation of a line in slope-intercept form that has a slope of -2 and a y-intercept of 3.
2. Factor the expression: $x^2 - 8x + 15$.
3. Solve the following quadratic equation using the quadratic formula:
$$4x^2 + 12x = 0$$
4. Simplify the following expression using exponential rules:
$$(4x^4y^2)^2$$

Use the following technical trade questions to test your understanding of pratical applications of Section 5–4.

1. **Electronics** The **period** of a waveform is the time it takes to complete one full cycle. A cycle is made up of a positive and negative alternation of voltage or current. The **frequency** of a waveform indicates the number of repetitions in one second. As an example, the electricity available at the wall outlet in your home is an alternating current (ac) sine wave that has a frequency of 60 hertz (Hz). This means that 60 cycles arrive every second, and so each cycle's period would be $\frac{1}{60}$ th of a second. Expressed mathematically:

$$\text{Frequency (in hertz)} = \frac{1}{\text{period (in seconds)}}$$

$$\text{and} \quad \text{Period} = \frac{1}{\text{frequency}}$$

a. Determine the period of the waveforms shown.

Sine wave — Frequency = 1000 hertz or 1 kilohertz

Square wave — Frequency = 24 kilohertz or 24 kHz

b. Determine the frequency of the waveforms shown.

Rectangular wave — Period = 1×10^{-3} seconds or 1 millisecond

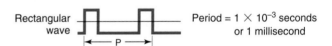

Sawtooth wave — Period = 560×10^{-6} seconds or 560 microseconds

2. **Sheet Metal** The formula for calculating the water pressure in pounds per square foot is:

$$P = 62.4 \times D$$

where P = pressure in pounds per square foot
D = depth of water in feet
62.4: weight of 1 cubic foot of water

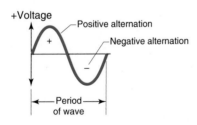

Frequency = 60 hertz

Period = $\dfrac{1}{\text{frequency}} = \dfrac{1}{60}$ = 16.67 milliseconds

Period = 16.67 milliseconds

Frequency = $\dfrac{1}{\text{period}} = \dfrac{1}{16.67 \text{ ms}}$ = 60 Hz

Water tank

a. Calculate the pressure applied to the bottom of a tank that holds 6 feet of water.
b. If a tank is constructed to withstand 1200 pounds per square foot of pressure, what is the maximum safe height of the tank?

3. **Health Care** The following formula is used to determine the maximum heart rate of a person during exercise, which depends on the person's age.

$$H_R = (-0.8 \times A) + 176$$

where H_R = heart rate in beats per minute (bpm)
A = age

a. A 40-year-old patient, when exercising on a treadmill, has a heart rate of 140 beats per minute. Is this value below the person's maximum target heart rate?
b. At what age should your heart rate not exceed 160 beats per minute?

5–5 RULES OF ALGEBRA—A SUMMARY

As a review of the lessons learned in this chapter, Table 5–1 summarizes the rules of algebra with many examples. Study every problem, since these will provide you with excellent practice for the applications that follow.

TABLE 5–1 Rules of Algebra—A Summary

Monomial (One Term): $2a$
Binomial (Two Terms): $2a + 3b$
Trinomial (Three Terms): $2a + 3b + c$
Polynomial (Any Number of Terms).
<u>Order of Operations:</u> Parentheses, Exponents, Multiplication, Division, Addition, Subtraction.
  (Memory Aid: Please, Excuse, My, Dear, Aunt, Sally)
<u>Basic Rules:</u>

$a + a = 2a$	$a - a = 0$	$a \cdot a = a^2$	$a \div a = 1$	$\sqrt{a^2} = a$
$a + 1 = a + 1$	$a - 1 = a - 1$	$a \cdot 1 = a$	$a \div 1 = a$	
$a + 0 = a$	$a - 0 = a$	$a \cdot 0 = 0$	$a \div 0 =$ Impossible	$0 \div a = 0$

<u>Parentheses First:</u> $(3 + 6) \cdot 2 = 9 \cdot 2 = 18$ $9 - (2 \cdot 3) = 9 - 6 = 3$ $6 + (9 - 7) = 6 + 2 = 8$
<u>No Parentheses</u>—Step 1: Multiplications and Divisions, left to right. $3 + 6 \cdot 4 = 3 + 24 = 27$
         Step 2: Additions and Subtractions, left to right. $4 \cdot 2 + 3 - 4 \div 2 = 8 + 3 - 2 = 9$
<u>Combine Only Like Terms:</u> $2x + x + y = 3x + y$ $3a + 4b - a = 2a + 4b$
<u>Commutative Property of Addition</u>—order of terms makes no difference: $a + b = b + a$
<u>Associative Property of Addition</u>—grouping of terms makes no difference: $(a + b) + c = a + (b + c)$
<u>Coefficients:</u> $x + x + x = 3x$ $x + y + x + y = 2x + 2y$
<u>Exponents:</u> $x \cdot x \cdot x = x^3$ $x \cdot y \cdot x \cdot y = x^2 \cdot y^2 = x^2 y^2$

$a^{-2} \cdot a^3 = a^{2+3} = a^5$ $\dfrac{a^3}{a^2} = a^{3-2} = a^1 = a$

$a^{-2} = \dfrac{1}{a^2}$ $(a^3)^2 = a^{3 \times 2} = a^6$

$\sqrt[3]{a^2} = (\sqrt[3]{a})^2$ $a^{\frac{2}{3}} = \sqrt[3]{a^2}$

<u>Commutative Property of Multiplication</u>—order of factors makes no difference: $a \cdot b = b \cdot a$
<u>Associative Property of Multiplication</u>—grouping of factors makes no difference: $(ab)c = a(bc)$
<u>Distributive Property of Multiplication</u>—$a(b + c) = ab + ac$ $2(a - 3b) = 2 \cdot a - 2 \cdot 3b = 2a - 6b$
<u>Transposition</u>—Step 1: Ensure unknown is above fraction bar. $x \cdot 8 = 16$ (divide both sides by 8, $\div 8$)
         Step 2: Isolate unknown. $(x \cdot 8) \div 8 = (16) \div 8,$ $x = 2$

Unknown Above Fraction Bar

$(2 \cdot a) + 5 = 23 \quad (-5)$
$\quad 2 \cdot a = 18 \quad (\div 2)$
$\quad\quad a = 9$

Unknown Below Fraction Bar

$\dfrac{72}{x} = 12 \quad (\cdot x)$
$72 = 12 \cdot x \quad (\div 12)$
$6 = x$

Unknown on Both Sides

$(6 \cdot y) + 7 = y + 27 \quad (-7)$
$\quad\quad 6y = y + 20 \quad (-y)$
$\quad\quad 5y = 20 \quad (\div 5)$
$\quad\quad y = 4$

Combining Unknowns

$x + 2x + 4(x + 2x) = 100$
$x + 2x + 4x + 8x = 100$
$15x = 100$
$x = 6.67$

Positive and Negative Number Rules

Addition ⊕ $(+) + (+) = (+), (-) + (-) = (-)$
               $(+) + (-)$ or $(-) + (+) =$ Difference

Subtraction ⊖   Change sign of second number, and then add.

Multiplication ⊗ $(+) \cdot (+) = (+), (-) \cdot (-) = (+)$
               $(+) \cdot (-)$ or $(-) \cdot (+) = (-)$

Division ⊘ $(+) \div (+) = (+), (-) \div (-) = (+)$
               $(+) \div (-)$ or $(-) \div (+) = (-)$

Positive and Negative Numbers

Addition ⊕ $2x + 5x = 7x \quad -6a + (+3a) = -3a$

Subtraction ⊖ $-3y^2 - (-2y^2) = -3y^2 + (+2y^2) = -y^2$

Multiplication ⊗ $-3(6a) = -3 \cdot (+6a) = -18a$

Division ⊘ $14a \div (-7) = -2a$

Adding Polynomials: $(4a^2 + 6b^2 + x - 2) + (7a^2 - b^2 + 3)$

$\quad\quad 4a^2 + 6b^2 + x - 2$
$+ \quad 7a^2 - b^2 \quad\quad\quad + 3$
$\overline{\quad 11a^2 + 5b^2 + x + 1 \quad}$

Subtracting Polynomials: $(6x^2 + y^2 - 4) - (-4x^2 + 2y^2 + 6)$

$\quad\quad 6x^2 + y^2 - 4 \quad\quad\quad\quad\quad\quad 6x^2 + y^2 - 4$
$- \quad -4x^2 + 2y^2 + 6 \quad$ Rule: Add $\quad + \quad 4x^2 - 2y^2 - 6$
$\quad\quad\quad\quad\quad\quad\quad\quad\quad$ the Opposite → $\overline{\quad 10x^2 - y^2 - 10 \quad}$

Multiplying Monomials, Binomials

$x^2 \cdot x^3 \cdot x = x \cdot x \cdot x \cdot x \cdot x \cdot x = x^{2+3+1} = x^6$
$(x^3)^2 = (x \cdot x \cdot x) \cdot (x \cdot x \cdot x) = x^{3 \cdot 2} = x^6$
$2(a + 2b) = 2a + 4b$

$4x^2 \cdot x^3 = 4 \cdot x \cdot x \cdot x \cdot x \cdot x = 4x^{2+3} = 4x^5$
$(-6y)^2 \cdot y^3 = (-6)^2 \cdot y^2 \cdot y^3 = 36y^5$
$-2(2a - 3b) = (-2 \cdot 2a) - (-2 \cdot 3b) = -4a - (-6b) = p - 4a + 6b$

Multiplying Polynomials

$\quad\quad x^2 + 2x - 1 \quad\quad (x^2 + 2x - 1) \cdot (x - 3) = ?$
$\times \quad\quad\quad x - 3$
$\overline{\quad -3x^2 - 6x + 3 \quad} \leftarrow -3(x^2 + 2x - 1)$
$\quad x^3 + 2x^2 - x \quad\quad \leftarrow x(x^2 - 2x - 1)$
$\overline{\quad x^3 - x^2 - 7x + 3 \quad}$

$\quad\quad a - 3 \quad\quad (a - 3)^2 = ?$
$\times \quad a - 3$
$\overline{\quad -3a + 9 \quad} \leftarrow -3(a - 3)$
$\quad a^2 - 3a \quad \leftarrow a(a - 3)$
$\overline{\quad a^2 - 6a + 9 \quad}$

Dividing Polynomials

$\dfrac{a^4}{a} = \dfrac{{}^1\cancel{a} \cdot a \cdot a \cdot a}{{}^1\cancel{a}} = a^3$

$\dfrac{2ab - 4a^2b^2}{2a} = \dfrac{2ab}{2a} - \dfrac{4a^2b^2}{2a} = b - 2ab^2$

$\dfrac{-6ab^3}{2ab} = -\dfrac{{}^{-3}\cancel{6} \cdot \cancel{a} \cdot \cancel{b} \cdot b \cdot b}{{}_1\cancel{2} \cdot \cancel{a} \cdot \cancel{b}} = -3b^2$

$\dfrac{2a^2 + 6a + 8ab}{2a} = \dfrac{2a^2}{2a} + \dfrac{6a}{2a} + \dfrac{8ab}{2a} = a + 3 + 4b$

Linear Equations (literals raised to first power):

    Standard Form (STF): $Ax + By = C$ Slope $= \dfrac{\Delta y}{\Delta x}$ ("/" slope $= +$, "\" slope $= -$)

    Slope-Intercept Form (SIF): $y = mx + b$

    Point-Slope Form (PSF): $y - y_1 = m(x - x_1)$

Factoring Patterns ($x^2 + bx + c = (x + m)(x + n)$):

         ▨ $+$ ▨ $+$ ▨ $= (x + m)(x + n)$

         ▨ $+$ ▨ $-$ ▨ $= (x + m)(x - n)$ or $(x - m)(x + n)$

         ▨ $-$ ▨ $+$ ▨   or   ▨ $-$ ▨ $-$ ▨ $= (x - m)(x - n)$

     Difference of Perfect Squares: $x^2 - y^2 = (x - y)(x + y)$

     Sum of Perfect Cubes: $x^3 + y^3 = (x + y)(x^2 - xy + y^2)$

     Difference of Perfect Cubes: $x^3 - y^3 = (x - y)(x^2 + xy + y^2)$

Quadratic Equations (literals raised to second power):

    Standard Form (STF): $ax^2 + bx + c = 0$

    Quadratic Formula: $x = \dfrac{-b \pm \sqrt{b^2 - 4ac}}{2a}$

SELF-TEST EVALUATION POINT FOR SECTION 5–5

Now that you have completed this section, you should be able to:

■ **Objective 12.** *Summarize the terms, rules, properties, and mathematical operations of algebra.*

Use the following questions to test your understanding of Section 5–5.

1. What is the order of operations?
2. Give an example of a:
   a. Monomial
   b. Binomial
   c. Trinomial
   d. Polynomial
3. Determine the following:
   a. $a \div 1 = ?$
   b. $a + a = ?$
   c. $a - a = ?$
   d. $a \times a = ?$
   e. $3x + b - 2x = ?$
4. Determine the unknown:
   a. $16 + a = 5 \times a, a = ?$
   b. $\dfrac{42}{b} = 3 \times 7$

REVIEW QUESTIONS

Multiple Choice Questions

1. If you perform exactly the same mathematical operation on both sides of an equation, the equality is preserved.
   a. True
   b. False

2. Are the two sides of the following equation, $(3 - 2) \times 16 = 32 \div 2$, equal?
   a. Yes
   b. No
   c. Only if the right half is doubled
   d. Only if the left half is doubled

3. $x + x =$ _____, $x \times x =$ _____, $x \div x =$ _____, and $x - x =$ _____.
   a. $x^2, 2x, 1, 0$
   b. $2x, \sqrt{x}, 0, 1$
   c. $1x, 2x, 1, 0$
   d. $2x, x^2, 1, 0$

4. What is the value of the unknown in the following equation: $3a + 9 = 2a + 10$?
   a. 2
   b. 10
   c. 1
   d. 3

5. Calculate the value x for the following equation: $x + y^2 = y^2 + 9$.
   a. 3
   b. 6
   c. 9
   d. 12

6. What is the value of the unknown, s, in the following equation: $\sqrt{s^2} + 14 = 26$?
   a. 12
   b. 26
   c. 4
   d. 5

7. What is the correct transposition for the following formula: $a = \dfrac{b - c}{x}, x = ?$
   a. $x = \dfrac{b - a}{c}$
   b. $x = (b - a) \times c$
   c. $x = \dfrac{b - c}{a}$
   d. $x = x(b - c)$

8. If $V = I \times R$ and $P = V \times I$, develop a formula for P when I and R are known.
   a. $P = \dfrac{I}{R}$
   b. $P = I^2 \times R^2$
   c. $P = I^2 \times R$
   d. $P = \sqrt{\dfrac{I}{R}}$

9. If $E = M \times C^2, M = ?$ and $C = ?$
   a. $\sqrt{\dfrac{E}{C}}, \dfrac{E^2}{M}$
   b. $\dfrac{E}{C^2}, \sqrt{\dfrac{E}{M}}$
   c. $E \times C^2, \sqrt{C^2 \times E}$
   d. $\dfrac{C^2}{E}, \sqrt{\dfrac{M}{E}}$

10. If $P = \dfrac{V^2}{R}$, transpose to obtain a formula to solve for V.
    a. $V = \sqrt{P \times R}$
    b. $V = \dfrac{P^2}{R}$
    c. $V = P \times R^2$
    d. $V = \sqrt{P} \times R$

Communication Skill Questions

11. Define the following terms: (5–1)
    a. Algebra
    b. Formula
    c. Equation
    d. Literal number
12. Why is it important to treat both sides of an equation equally? (5–1)
13. Describe how the following three factors can be combined in a formula: distance (d), velocity (v), and time (t). (5–2)
14. What two steps must be followed to transpose an equation? (5–2)
15. Describe the factoring process. (5–2)
16. Define the terms *proportional* and *inversely proportional*. (5–2)
17. Explain how current (I), voltage (V), and resistance (R) can be combined in a formula. (5–2)
18. Transpose the formula from question 22 to solve for I, V, and R. (5–2)
19. If $P = V \times I$, describe how you would solve for V and I. (5–2)
20. Define substitution. (5–3)
21. If $\pi = C/d$, and area $= \frac{1}{2} \times (C \times r)$, describe how substitution and transposition can be employed to obtain a formula for area when only the radius of a circle is known. (5–3)
22. Explain how the parentheses can be removed from the following examples: (5–2)
    a. $4(3b - 1b)$
    b. $2(3x \times 2x)$
23. If $a = b \times c$, explain how you would solve for b and c. (5–2)
24. Describe the relationship between square and square root. (5–2)
25. Why is formula manipulation important? (5–1)

Practice Problems

26. Calculate the result of the following arithmetic operations involving literal numbers.
    a. $x + x + x = ?$
    b. $5x + 2x = ?$
    c. $y - y = ?$
    d. $2x - x = ?$
    e. $y \times y = ?$
    f. $2x \times 4x = ?$
    g. $4y \times 3y^2 = ?$
    h. $a \times a = ?$
    i. $2y \div y = ?$
    j. $6b \div 3b = ?$
    k. $\sqrt{a^2} = ?$
    l. $(x^3)^2 = ?$

27. Transpose the following equations to determine the unknown value.
    a. $4x = 11$
    b. $6a + 4a = 70$
    c. $5b - 4b = \dfrac{7.5}{1.25}$
    d. $\dfrac{2z \times 3z}{4.5} = 2z$

28. Apply transposition and substitution to Newton's second law of motion.
    a. $F = ma$ [Force (F) = mass (m) × acceleration (a)]
       If $F = m \times a$, $m = ?$ and $a = ?$
    b. If acceleration (a) equals change in speed (Δv) divided by time (t), how would the $F = ma$ formula appear if acceleration (a) was substituted for $\dfrac{\Delta v}{t}$?

29. a. Using the formula power (P) = voltage (V) × current (I), calculate the value of electric current through a hair dryer that is rated as follows: power = 1500 watts and voltage = 120 volts.

    $$P = V \times I$$

    where P = power in watts
    V = voltage in volts
    I = current in amperes

    b. Using Ohm's formula voltage (V) = current (I) × resistance (R), calculate the resistance or opposition to current flow offered by the hairdryer in Question 14a.

30. The following formula is used to calculate the total resistance (R_T) in a parallel circuit containing two parallel resistances (R_1 and R_2).

    $$R_T = \dfrac{1}{\dfrac{1}{R_1} + \dfrac{1}{R_2}}$$

    a. Calculate R_T (total resistance) if the resistance of $R_1 = 6$ ohms and the resistance of $R_2 = 12$ ohms.
    b. Indicate the calculator sequence used for part (a).
    c. If R_2 and R_T were known and we needed to calculate the value of R_1, we would have to transpose the total parallel resistance formula to solve for R_1. The following steps show how to transpose the total resistance formula to calculate R_1 when R_T and R_2 are known. Describe what has occurred in each of the steps.

    $R_T = \dfrac{1}{\dfrac{1}{R_1} + \dfrac{1}{R_2}}$ (Original formula)

    1. $R_T \times \left(\dfrac{1}{R_1} + \dfrac{1}{R_2}\right) = \dfrac{1}{\dfrac{1}{R_1} + \dfrac{1}{R_2}} \times \left(\dfrac{1}{R_1} + \dfrac{1}{R_2}\right)$

    $R_T \times \left(\dfrac{1}{R_1} + \dfrac{1}{R_2}\right) = 1$

    2. $\dfrac{R_T \times \left(\dfrac{1}{R_1} + \dfrac{1}{R_2}\right)}{R_T} = \dfrac{1}{R_T}$

    $\dfrac{1}{R_1} + \dfrac{1}{R_2} = \dfrac{1}{R_T}$

or $\dfrac{1}{R_T} = \dfrac{1}{R_1} + \dfrac{1}{R_2}$

3. $\dfrac{1}{R_T} - \dfrac{1}{R_2} = \dfrac{1}{R_1} + \dfrac{1}{\cancel{R_2}} - \dfrac{1}{\cancel{R_2}}$

$\dfrac{1}{R_T} - \dfrac{1}{R_2} = \dfrac{1}{R_1}$

or $\dfrac{1}{R_1} = \dfrac{1}{R_T} - \dfrac{1}{R_2}$

4. $\dfrac{1}{R_1} = \dfrac{1}{R_T} - \dfrac{1}{R_2}$

$\dfrac{1}{R_T} - \dfrac{1}{R_2} = $ *lowest common denominator (LCD)*

$\dfrac{1}{R_T} - \dfrac{1}{R_2} = \dfrac{1}{R_T \times R_2}$ (LCD) method 1: fractions
$R_T \times R_2 = R_T \times R_2$

$\dfrac{1}{R_T} - \dfrac{1}{R_2} = \dfrac{R_2 - R_T}{R_T \times R_2}$ *Steps:*
$(R_T \times R_2) \div R_T = R_2$,
$R_2 \times 1 = R_2$

$\dfrac{1}{R_T} - \dfrac{1}{R_2} = \dfrac{R_2 - R_T}{R_T \times R_2}$

$\dfrac{1}{R_1} = \dfrac{R_2 - R_T}{R_T \times R_2}$ $(R_T \times R_2) \div R_2 = R_T$,
$R_T \times 1 = R_T$

5. $\dfrac{1}{\cancel{R_1}} \times \cancel{R_1} = \dfrac{R_2 - R_T}{R_T \times R_2} \times R_1$

$1 = \dfrac{R_2 - R_T}{R_T \times R_2} \times R_1$

6. $1 \times R_T \times R_2 = \dfrac{R_2 - R_T}{\cancel{R_T} \times \cancel{R_2}} \times R_1 \times \cancel{R_T} \times \cancel{R_2}$

$R_T \times R_2 = (R_2 - R_T) \times R_1$

7. $\dfrac{R_T \times R_2}{R_2 - R_T} = \dfrac{(\cancel{R_2 - R_T}) \times R_1}{\cancel{R_2 - R_T}}$

$\dfrac{R_T \times R_2}{R_2 - R_T} = R_1$ or $R_1 = \dfrac{R_T \times R_2}{R_2 - R_T}$

d. Using the values from part (a), double-check the new formula.
e. If we wanted to obtain a formula for calculating R_2 when both R_T and R_1 are known, which of the steps would we change, and in what way?
f. Give the formula for calculating R_2 when both R_T and R_1 are known.
g. Using the values from part (a), double-check this new formula.

Math Application Practice Problems

31. **Personal Finance** A car payment of $373 leaves a bank balance of $1269. How much was in the account before the payment?

32. **Aerospace** A jet engine under test generates a sound intensity of 115 decibels. If ear plugs reduce the noise level by 37 decibels, what sound level will be heard by the technician?

33. **Automotive** An oil change at a gas station costs $16.38 less than at the dealer. If a tune-up and oil change costs $98, and the tune-up costs $48, how much will the oil change cost at the gas station?

34. **Electrical** A circuit breaker is rated at 15 amps. The current drain at present is 12.6 amps. How much more current can be pulled by the circuit before the circuit breaker trips?

35. **Accounting** A spreadsheet requires 112 cells and will use columns A through G. How many rows will be needed?

36. **Plumbing** One length of hot water pipe is 3 feet longer than the other. If their combined length is 27 feet, how long is the shorter pipe?

37. **Manufacturing** When manufacturing cell phone integrated circuits, 18 out of 43 are defective. What percentage of failure is this? (Round to the nearest one percent.)

38. **Painting** A tarpaulin can cover an area of 48 square feet. If one of its sides is 8 feet long, how long is the other side?

39. **Health Care** A paramedic performs 25 compressions to 10 breaths on an adult patient who does not have a pulse. What is the compression-to-breath ratio, and how many breaths would be needed if 40 compressions were administered?

40. **Electronics** A computer can perform 432 floating-point operations in 3.6 seconds. How many floating-point operations (FLOPS) can it perform in 5 seconds?

Web Site Questions

Go to the Web site http://www.prenhall.com/cook, select the textbook *Mathematics for the Technical Trades*, select this chapter, and then follow the instructions when answering the multiple-choice practice problems.

Geometry and Trigonometry

Finding the Question to the Answer!

More than 350 years ago in 1637, Pierre de Fermat, a French mathematician and physicist, wrote an apparently simple theorem in the margins of a book. He also added that he had discovered marvelous proof of it but lacked enough space to include it in the margin. Later that night he died and took with him a mystery that mathematicians have been trying to solve ever since. Many of the brightest minds in mathematics have struggled to find the proof, and many have concluded that Fermat must have been mistaken despite his considerable mathematical ability.

Fermat's last theorem, as it has been appropriately named, has to do with equations of the form x to the nth power, plus y to the nth power = z to the nth power.

$$x^n + y^n = z^n$$

A power of 2 (square) is familiar as the Pythagorean theorem, which states that the sum of the squares of the lengths of two sides of a right-angle triangle is equal to the square of the length of the other side:

$$c^2 = a^2 + b^2$$

For example,

$$5^2 = 3^2 + 4^2$$
$$25 = 9 + 16$$

Fermat's last theorem seemed not to work because, for example,

$$? = 3^3 + 4^3$$
$$91 = 27 + 64$$

and there is not a whole number that can be cubed to equal 91.

Now, after thousands of claims of success that proved untrue, mathematicians say the daunting challenge, perhaps the most famous of unsolved mathematical problems, has been surmounted. The conqueror is Andrew Wiles, 40, a British mathematician who works at Princeton University. Wiles announced the result at the last of three lectures at Cambridge University in England. Within a few minutes of the conclusion of his final lecture, computer mail messages were winging around the world, sending mathematicians into a frenzy. It had apparently taken 7 years for Wiles to solve the problem. He continued a chain of ideas that was begun in 1954 by several other mathematicians.

After being criticized by many in the eighteenth and nineteenth centuries and ridiculed in the twentieth century by mathematicians with supercomputers, Fermat was elevated finally in 1993 to the status of "a genius before his time."

6

Outline and Objectives

VIGNETTE: FINDING THE QUESTION TO THE ANSWER!

INTRODUCTION

6–1 BASIC GEOMETRY TERMS

Objective 1: Describe the purpose of geometry, and define many of the basic geometric terms.

Objective 2: Demonstrate how to use a protractor to measure and draw an angle.

6–2 PLANE FIGURES

Objective 3: Define the term *plane figure,* and identify the different types of polygons.

Objective 4: List the formulas for determining the perimeter and area of plane figures.

6–2–1 Quadrilaterals

Objective 5: Explain how the Pythagorean theorem relates to right-angle triangles.

Objective 6: Determine the length of one side of a right-angle triangle when the lengths of the other two sides are known.

Objective 7: Describe how vectors are used to represent the magnitude and direction of physical quantities and how they are arranged in a vector diagram.

Objective 8: Explain how the Pythagorean theorem can be applied to a vector diagram to calculate the magnitude of a resultant vector.

6–2–2 Right-Angle Triangles
6–2–3 Vectors and Vector Diagrams
6–2–4 Other Triangles
6–2–5 Circles
6–2–6 Other Polygons

6–3 TRIGONOMETRY

Objective 9: Define the trigonometric terms:
    a. Opposite
    b. Adjacent
    c. Hypotenuse
    d. Theta
    e. Sine
    f. Cosine
    g. Tangent

Objective 10: Demonstrate how the sine, cosine, and tangent trigonometric functions can be used to calculate:
    a. The length of an unknown side of a right-angle triangle if the length of another side and the angle theta are known
    b. The angle theta of a right-angle triangle if the lengths of two sides of the triangle are known

6–3–1 Sine of Theta
6–3–2 Cosine of Theta
6–3–3 Tangent of Theta
6–3–4 Summary

6–4 SOLID FIGURES

Objective 11: Define the term *solid figure,* and identify the different types.

Objective 12: List the formulas for determining the lateral surface area, total surface area, and volume of solid figures.

6–4–1 Prisms
6–4–2 Cylinders
6–4–3 Spheres
6–4–4 Pyramids and Cones

Introduction

Geometry is one of the oldest branches of mathematics and deals with the measurement, properties, and relationships of points, lines, angles, plane figures or surfaces, and solid figures. The practical attributes of geometry were used by the ancient Egyptians to build the pyramids, and today, it is widely used in almost every branch of engineering technology. Unlike algebra, which deals with unknowns that are normally represented by a letter, geometry deals with a physical point, line, surface, or figure.

The first section of this chapter deals with the basic elements of geometry, the second section covers plane figures, the third trigonometry, and the final section examines solid figures.

6–1 BASIC GEOMETRIC TERMS

Figure 6–1(a) shows some examples of line segments. In the first example you can see that a line segment exists between ending points A and B. An abbreviation is generally used to represent a line segment, in which a bar is placed over the two end-point letters, and so "line segment AB" could be written as $\overline{AB}$.

Figure 6–1(a) also shows an example of a vertical line that extends straight up and down, a horizontal line that extends from left to right, and two parallel lines that extend in the same direction while remaining at the same distance from each other.

In Figure 6–1(b) an angle is formed between two intersecting line segments or lines, which are also called **sides.** The common point at which they connect is called the **vertex.** The angle symbol $\angle$ is generally used as an abbreviation, and so "angle ABC" could be written as $\angle ABC$. The unit of measurement for the opening between the sides, or angle, is the degree (symbolized " ° ").

Figure 6–1(c) shows how a protractor can be used to measure angles. An image of a protractor has been printed in the back of this text and can, if copied, be used for measuring and drawing angles. In this example, we are using the protractor to measure $\angle ABC$, from Figure 6–1(b). The three-step procedure for measuring an angle using a protractor is as follows:

Sides
The lines that form an angle.

Vertex
A point where two lines meet.

MEASURING AN ANGLE WITH A PROTRACTOR

Step 1: Place protractor's center mark on vertex.
Step 2: Rotate protractor until zero-degree mark is on one line.
Step 3: Read the measure in degrees where the line intersects the protractor's scale.

You may have noticed that the protractor has a reversing upper and lower scale, so that an angle can be read either clockwise or counterclockwise. Thus if you were to measure the angle in Figure 6–1(b), you would use the lower scale of the protractor as shown in Figure 6–1(c), moving counterclockwise from 0°. On the other hand, if the angle was in a reverse direction as shown with the 90° angle in Figure 6–1(d), you would use the upper scale of the protractor moving clockwise from 0°.

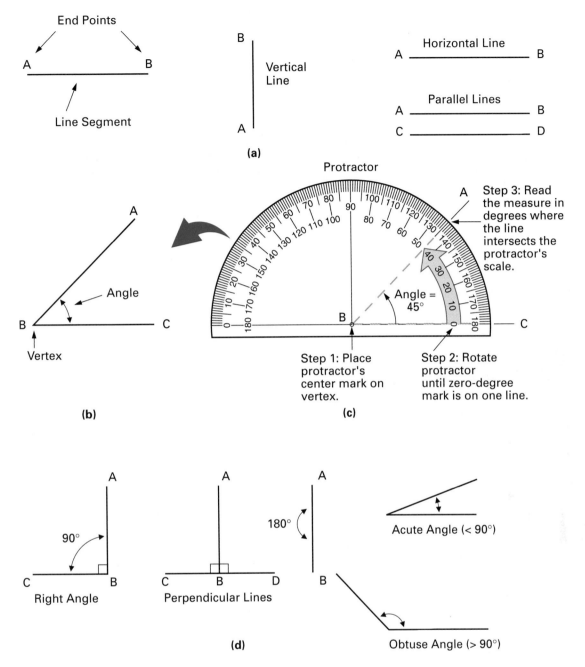

FIGURE 6–1 Basic Geometric Terms.

The protractor can also be used to draw an angle. The three-step procedure for drawing an angle using a protractor is as follows:

DRAWING AN ANGLE WITH A PROTRACTOR

Step 1: Draw a straight line on the paper representing one side of the angle.

Step 2: Place the protractor's center mark on the end of the line that is to serve as the angle's vertex.

Step 3: Draw a dot on the paper next to the protractor's scale equal to the measure of the desired angle, and then draw a line connecting this dot to the vertex.

Acute Angle
An angle that measures less than 90°.

Obtuse Angle
An angle that measures greater than 90°.

Figure 6–1(d) shows a few more geometric basics. A right angle is one of 90° and can be symbolized with a small square in the corner. Two lines are perpendicular to each other when the point at which they intersect forms right angles. The angle measure of a straight angle is 180°. The terms **acute angle** and **obtuse angle** describe angles that are relative to a right angle (90°). An acute angle is one that is less than a right angle (< 90°), while an obtuse angle is one that is greater than a right angle (> 90°).

SELF-TEST EVALUATION POINT FOR SECTION 6–1

Now that you have completed this section, you should be able to:

■ *Objective 1.* Describe the purpose of geometry, and define many of the basic geometric terms.

■ *Objective 2.* Demonstrate how to use a protractor to measure and draw an angle.

Use the following questions to test your understanding of Section 6–1:

1. Using a protractor, measure:
   a. ∠AXB   d. ∠AXE   g. ∠AXH   j. ∠IXG
   b. ∠AXC   e. ∠AXF   h. ∠AXI   k. ∠IXF
   c. ∠AXD   f. ∠AXG   i. ∠IXH   l. ∠IXE

2. Which of the angles in the previous question are right angles, obtuse angles, and acute angles?

3. Draw:
   a. A vertical line
   b. A horizontal line
   c. Three parallel lines
   d. A right angle
   e. An angle of 35°
   f. An angle of 172°
   g. An obtuse angle
   h. An acute angle

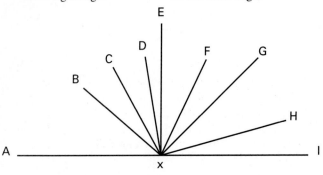

Use the following technical trade questions to test your understanding of practical applications of Section 6–1.

1. **Construction** Referring to the roof in the drawing, determine the angle at which the rafters should be set.

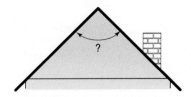

2. **Electrical** What are the angles of the elbows being used to connect conduit:
   a. A to B
   b. C to D

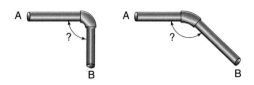

3. **Machining** The technical drawing details a part to be fabricated. What is the angle of the flare?

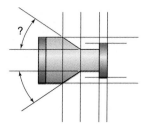

4. **Carpentry** A carpenter has been contracted to construct an enclosure for the pool equipment shown in the figure. At what angle will the cover be cut?

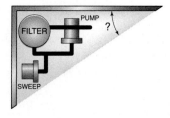

6–2 PLANE FIGURES

A plane figure is a perfectly flat surface, such as a tabletop, floor, wall, or windowpane. In this section we will examine the details of several flat plane surfaces, which, as you will discover, have only two dimensions—length and width.

6–2–1 *Quadrilaterals*

A **polygon** is a plane geometric figure that is made up of three or more line segments. Quadrilaterals, such as the square, rectangle, rhombus, parallelogram, and trapezoid, are all examples of four-sided polygons.

The information given in Figure 6–2 summarizes the facts and formulas for the square, which has four right angles and four equal length sides, and is a **parallelogram** (which means it has two sets of parallel sides). The box in Figure 6–2 lists the formulas for calculating a square's perimeter, area, and side length.

Figures 6–3, 6–4, 6–5, and 6–6 summarize the facts and formulas for other quadrilaterals. In addition, these diagrams include, when necessary, a visual breakdown of the formulas so that it is easier to see how the formula was derived.

Polygon
A plane geometry figure bounded by three or more line segments.

Parallelogram
A quadrilateral in which both pairs of opposite sides are parallel.

6–2–2 *Right-Angle Triangles*

The **right-angle triangle,** or right triangle, shown in Figure 6–7(a) (p. 176), has three sides and three corners. Its distinguishing feature, however, is that two of the sides of this triangle are at right angles (at 90°) to each other. The small square box within the triangle is placed in the corner to show that sides *A* and *B* are *square* or at *right angles* to each other.

If you study this right triangle you may notice two interesting facts about the relative lengths of sides *A*, *B*, and *C*. These observations are as follows.

1. Side *C* is always longer than side *A* or side *B*.
2. The total length of side *A* and side *B* is always longer than side *C*.

Right-Angle Triangle
A triangle that contains one right or 90° angle.

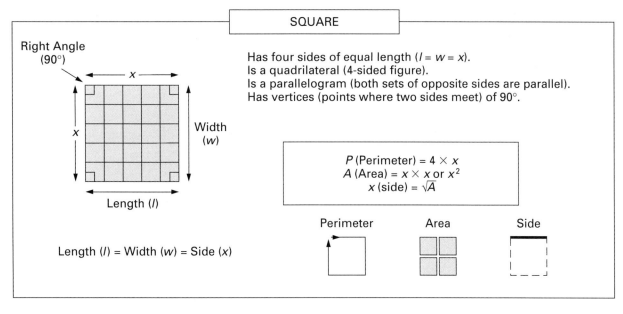

FIGURE 6–2 The Square.

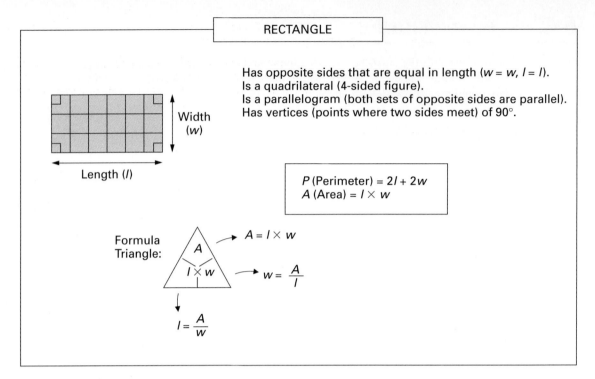

FIGURE 6–3 The Rectangle.

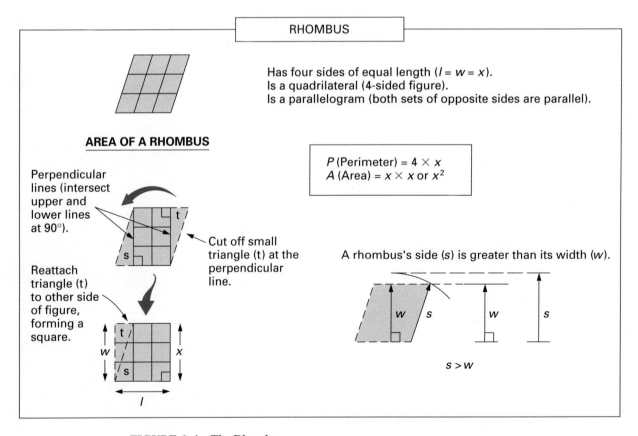

FIGURE 6–4 The Rhombus.

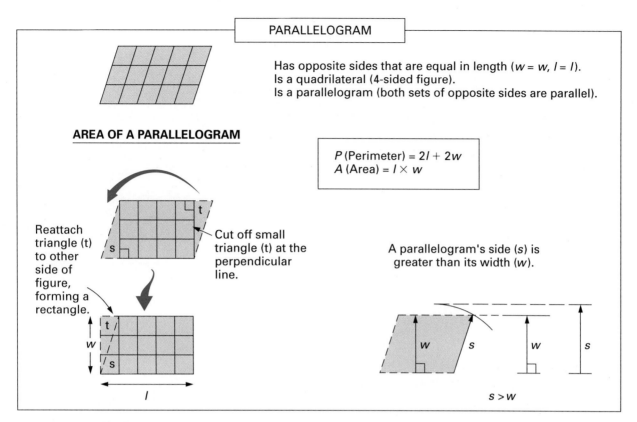

FIGURE 6–5 The Parallelogram.

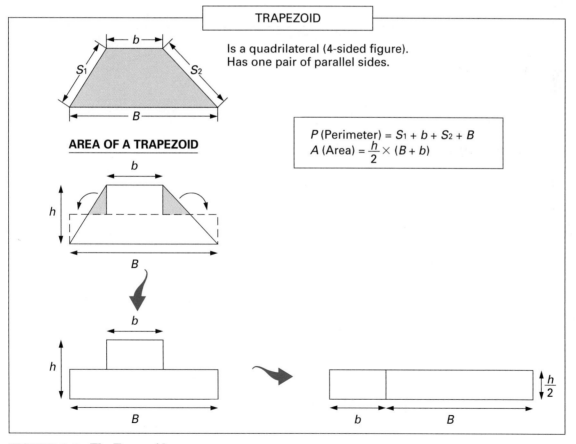

FIGURE 6–6 The Trapezoid.

175

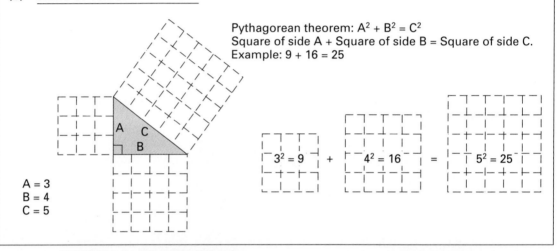

FIGURE 6–7 The Right Triangle.

The right triangle in Figure 6–7(b) has been drawn to scale to demonstrate another interesting fact about this triangle. If side *A* were to equal 3 cm and side *B* were to equal 4 cm, side *C* would equal 5 cm. This demonstrates a basic relationship among the three sides and accounts for why this right triangle is sometimes referred to as a *3, 4, 5 triangle.* As long as the relative lengths remain the same, it makes no difference whether the sides are 3 cm, 4 cm, and 5 cm or 30 km, 40 km, and 50 km. Unfortunately, in most applications our lengths of *A*, *B*, and *C* will not work out as easily as 3, 4, 5.

1. Pythagoras's Theorem

When a relationship exists among three quantities, we can develop a formula to calculate an unknown when two of the quantities are known. It was Pythagoras who first developed this basic formula or equation (known as the **Pythagorean theorem**), which states that *the square of the length of the hypotenuse (side C) of a right triangle equals the sum of the squares of the lengths of the other two sides:*

$$C^2 = A^2 + B^2 \qquad \begin{pmatrix} 5^2 = 3^2 + 4^2 \\ 25 = 9 + 16 \end{pmatrix}$$

By using the rules of algebra, we can transpose this formula to derive formulas for sides *C*, *B*, and *A*.

Pythagorean Theorem
A theorem in geometry: The square of the length of the hypotenuse of a right triangle equals the sum of the squares of the lengths of the other two sides.

CHAPTER 6 / GEOMETRY AND TRIGONOMETRY

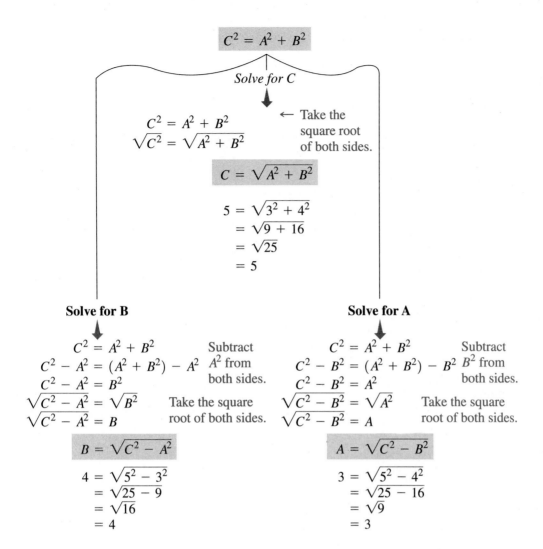

Let us now test these formulas with a few examples.

TECHNICAL TRADE APPLICATION: PAINTING

A 4-foot ladder has been placed in a position 2 feet from a wall. How far up the wall will the ladder reach?

■ *Solution:*

In this example, A is unknown, and therefore

$$A = \sqrt{C^2 - B^2}$$
$$= \sqrt{4^2 - 2^2}$$
$$= \sqrt{16 - 4}$$
$$= \sqrt{12}$$
$$= 3.46 \quad \text{or} \quad \approx 3.5 \text{ feet}$$

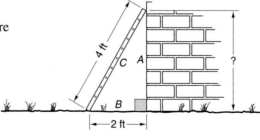

■ TECHNICAL TRADE APPLICATION: CONSTRUCTION

A 16-foot flagpole is casting a 12-foot shadow. What is the distance from the end of the shadow to the top of the flagpole?

■ *Solution:*

In this example, C is unknown, and therefore

$$\begin{aligned}C &= \sqrt{A^2 + B^2} \\ &= \sqrt{16^2 + 12^2} \\ &= \sqrt{256 + 144} \\ &= \sqrt{400} \\ &= 20 \text{ feet}\end{aligned}$$

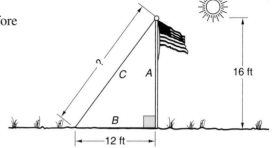

■ TECHNICAL TRADE APPLICATION: GEOLOGY

A 200-foot piece of string is stretched from the top of a 125-foot cliff to a point on the beach. What is the distance from this point to the cliff?

■ *Solution:*

In this example, B is unknown, and therefore

$$\begin{aligned}B &= \sqrt{C^2 - A^2} \\ &= \sqrt{200^2 - 125^2} \\ &= \sqrt{40{,}000 - 15{,}625} \\ &= \sqrt{24{,}375} \\ &= 156 \text{ feet}\end{aligned}$$

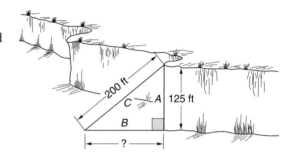

6-2-3 Vectors and Vector Diagrams

Vector or Phasor
A quantity that has magnitude and direction and that is commonly represented by a directed line segment whose length represents the magnitude and whose orientation in space represents the direction.

Vector Addition
Determination of the sum of two out-of-phase vectors using the Pythagorean theorem.

Resultant Vector
A vector derived from or resulting from two or more other vectors.

Vector Diagram
A graphic drawing using vectors that shows arrangement and relations.

A **vector or phasor** is an arrow used to represent the magnitude and direction of a quantity. Vectors are generally used to represent a physical quantity that has two properties. For example, Figure 6–8(a) shows a motorboat heading north at 12 miles per hour. Figure 6–8(b) shows how vector **A** could be used to represent the vessel's direction and speed. The size of the vector represents the speed of 12 miles per hour by being 12 centimeters long, and because we have made the top of the page north, vector **A** should point straight up so that it represents the vessel's direction. Referring to Figure 6–8(a) you can see that there is another factor that also needs to be considered, a 12 mile per hour easterly tide. This tide is represented in our vector diagram in Figure 6–8(b) by vector **B**, which is 12 centimeters long and pointing east. Since the motorboat is pushing north at 12 miles per hour and the tide is pushing east at 12 miles per hour, the resultant course will be northeast, as indicated by vector **C** in Figure 6–8(b). Vector **C** was determined by **vector addition** (as seen by the dashed lines) of vectors **A** and **B**. The result, vector **C,** is called the **resultant vector.** This resultant vector indicates the motorboat's course and speed. The direction or course, we can see, is northeast because vector **C** points in a direction midway between north and east. The speed of the motorboat, however, is indicated by the length or magnitude of vector **C.** This length, and therefore the motorboat's speed, can be calculated by using the Pythagorean theorem. Probably your next question is: How is the **vector diagram** in Figure 6–8(b) similar to a right-angle triangle? The answer is to redraw Figure 6–8(b), as shown in Figure 6–8(c). Because the dashed line x is equal in length to vector **A**, vector **A** can be put in the position of line x to form a right-angle triangle with vector **B** and vector **C.** Now that we have a right-angle

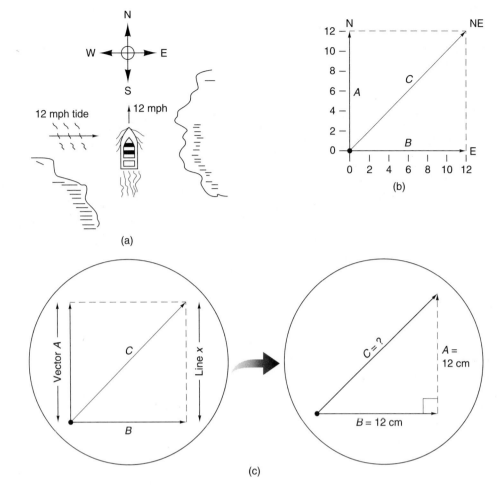

FIGURE 6–8 Vectors and Vector Diagrams.

triangle with two known lengths and one unknown length, we can calculate the unknown vector's length and therefore the motorboat's speed.

$$\begin{aligned} C &= \sqrt{A^2 + B^2} \\ &= \sqrt{12^2 + 12^2} \\ &= \sqrt{144 + 144} \\ &= \sqrt{288} = 16.97 \end{aligned}$$

The length of vector **C** is 16.97 cm, and because each 1 mile per hour of the motorboat was represented by 1 centimeter, the motorboat will travel at a speed of 16.97 miles per hour in a northeasterly direction.

■ TECHNICAL TRADE APPLICATION: ELECTRONICS

An electrical circuit has two elements that oppose current flow; one form of opposition is called *resistance* (symbolized by R), and the other form of opposition is called *reactance* (symbolized by X). These two oppositions are out of sync, or not in phase, with each other, as seen by the vector diagram in Figure 6–9(a), and therefore cannot simply be added. Using the Pythagorean theorem, vectorially add the two forms of opposition (R and X) to determine the total opposition to current flow, which is called *impedance* (symbolized by Z).

FIGURE 6–9 An Electrical Application of the Pythagorean Theorem.

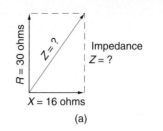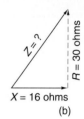

■ *Solution:*

First, redraw the vector diagram to produce a right-angle triangle, as shown in Figure 6–9(b). Then calculate Z:

$$Z = \sqrt{R^2 + X^2}$$
$$= \sqrt{30^2 + 16^2}$$
$$= \sqrt{900 + 256}$$
$$= \sqrt{1156}$$
$$= 34 \; \Omega$$

The total opposition or impedance (Z) to current flow is therefore 34 ohms.

6–2–4 Other Triangles

Figures 6–10, 6–11, and 6–12 summarize the facts and formulas for other triangle types.

SCALENE TRIANGLE

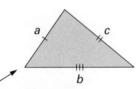

Is a triangle (3-sided figure).
Has no sides that are equal.

A different number of check marks on the sides of the triangle are used to indicate sides of different sizes.

P (Perimeter) $= a + b + c$
A (Area) $= \dfrac{b \times h}{2}$

If the length of the sides are known:

$A = \sqrt{s \times (s-a) \times (s-b) \times (s-c)}$

In which s (half perimeter) $= \dfrac{a + b + c}{2}$

AREA OF A SCALENE TRIANGLE

Scalene triangle occupies half the area of the rectangle below.

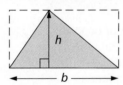

FIGURE 6–10 The Scalene Triangle.

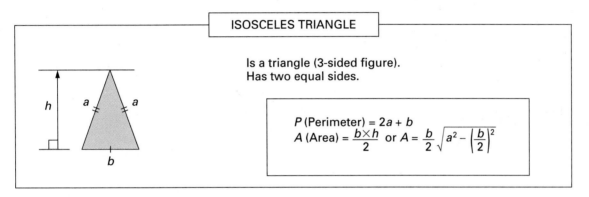

FIGURE 6–11 The Equilateral Triangle.

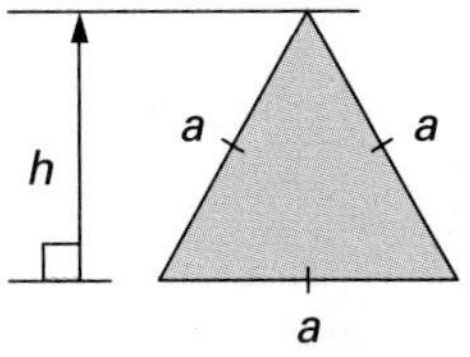

FIGURE 6–12 The Isosceles Triangle.

6–2–5 Circles

Details relating to the circle have been discussed in previous chapters. These facts and formulas are summarized in Figure 6–13.

6–2–6 Other Polygons

Figure 6–14(a) shows that if lines are drawn from the center of any polygon to each vertex, equivalent triangles are formed, which enables us to determine the polygon's area. In Figure 6–14(b), the formula given allows us to determine the number of degrees in the angles for each polygon type. From the examples shown we can see that each angle in an equilateral triangle is 60°; for a square each angle is 90°, as expected; for a pentagon each angle is 108°; a hexagon has six 120° angles; and an octagon has eight 135° vertices.

CIRCLE

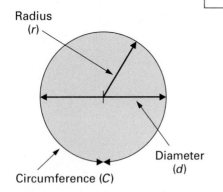

Is a perfectly round figure.
Every point is at an equal distance from the center.

Diameter $(d) = 2 \times r$
Circumference $(C) = \pi \times d$ or $c = \pi \times (2r)$
Area $(A) = \pi r^2$

Every circle's circumference is 3.14 times larger than its diameter. This constant is called pi (symbolized π).

$$\pi = \frac{\text{Circumference }(C)}{\text{Diameter }(d)} = 3.14$$

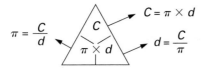

Since $d = 2r$, then:

$\pi = \dfrac{C}{2r}$ $C = \pi \times 2r$ $2r = \dfrac{C}{\pi}$

or $r = \dfrac{C}{2\pi}$

AREA OF A CIRCLE

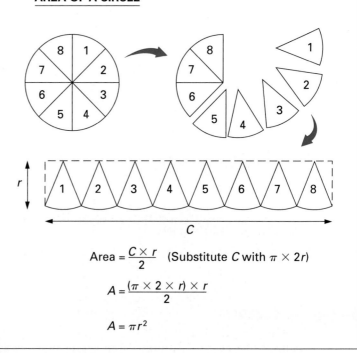

Area $= \dfrac{C \times r}{2}$ (Substitute C with $\pi \times 2r$)

$A = \dfrac{(\pi \times 2 \times r) \times r}{2}$

$A = \pi r^2$

FIGURE 6–13 The Circle.

OTHER POLYGONS

(a) Polygon Area

TRIANGLE
(3 sides)

SQUARE
(4 sides)

PENTAGON
(5 sides)

HEXAGON
(6 sides)

OCTAGON
(8 sides)

Equivalent triangles are created by drawing lines from the center of the polygon to each vertex.

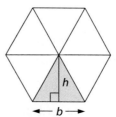

P (Perimeter) = # of Sides $\times$ b
A (Area) = $\frac{b \times h}{2}$ $\times$ # of Sides

(b) Number of Degrees in Each Vertex of a Polygon

$$x \text{ (Degrees in Vertex)} = \frac{180° \times (\# \text{ of Sides} - 2)}{\# \text{ of Sides}}$$

TRIANGLE: (Equilateral) $x = \frac{180° \times (3-2)}{3} = \frac{180°}{3} = 60°$ $x = 60°$ $3x = 180°$

SQUARE: $x = \frac{180° \times (4-2)}{4} = \frac{360°}{4} = 90°$ $x = 90°$ $4x = 360°$

PENTAGON: $x = \frac{180° \times (5-2)}{5} = \frac{540°}{5} = 108°$ $x = 108°$ $5x = 540°$

HEXAGON: $x = \frac{180° \times (6-2)}{6} = \frac{720°}{6} = 120°$ $x = 120°$ $6x = 720°$

OCTAGON: $x = \frac{180° \times (8-2)}{8} = \frac{1,080°}{8} = 135°$ $x = 135°$ $8x = 1,080°$

FIGURE 6–14 Other Polygons.

SELF-TEST EVALUATION POINT FOR SECTION 6–2

Now that you have completed this section, you should be able to:

■ *Objective 3.* Define the term plane figure, and identify the different types of polygons.

■ *Objective 4.* List the formulas for determining the perimeter and area of plane figures.

■ *Objective 5.* Explain how the Pythagorean theorem relates to right-angle triangles.

■ *Objective 6.* Determine the length of one side of a right-angle triangle when the lengths of the other two sides are known.

■ *Objective 7.* Describe how vectors are used to represent the magnitude and direction of physical quantities and how they are arranged in a vector diagram.

■ *Objective 8.* Explain how the Pythagorean theorem can be applied to a vector diagram to calculate the magnitude of a resultant vector.

Use the following questions to test your understanding of Section 6–2:

1. Define the terms:
   a. Plane figure
   b. Polygon

2. Sketch the following and list the formulas for perimeter and area:
   a. Square
   b. Rectangle
   c. Rhombus
   d. Parallelogram
   e. Trapezoid
   f. Right triangle
   g. Scalene triangle
   h. Equilateral triangle
   i. Isosceles triangle
   j. Circle

3. Calculate the lengths of the unknown sides in the following triangles.

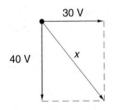

4. Calculate the magnitude of the resultant vectors in the following vector diagrams.

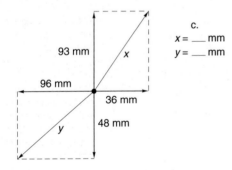

a. $x =$ ___ V
b. $x =$ ___ W
c. $x =$ ___ mm
   $y =$ ___ mm

Use the following technical trade questions to test your understanding of practical applications of Section 6–2.

1. **Electronics** The capacitor shown has two same-size metal foil plates that are separated by a nonconductive insulator called a *dielectric*. Unwrapped, the foil strips measure 6.2 centimeters by 1.3 centimeters. What is the capacitor's plate area?

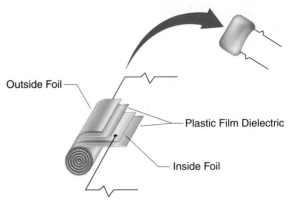

2. **Construction** How many packs of tongue-and-groove hardwood flooring would be needed to cover an 18-foot by 12-foot room? A pack can cover 16 square feet, and you should allow 25% extra for end and side matching.

3. **Carpentry** Calculate the area of the patio deck shown. Also determine the length of side C for a railing.

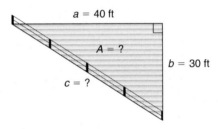

4. **Machining** Find the area of each of the following parts.

a.

b.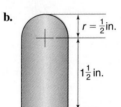

c.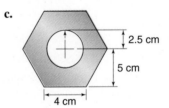

d.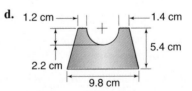

FIGURE 6–15 Names Given to the Sides of a Right Triangle.

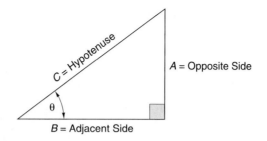

6–3 TRIGONOMETRY

In the previous section you saw how Pythagoras's theorem could be used to find the length of one side of a right triangle when the lengths of the other two sides are known. **Trigonometry** is the study of the properties of triangles, and in this section we will examine the right-angle triangle in a little more detail, and see how angles and sides are related.

Until this point we have called the three sides of our right triangle A, B, and C. Figure 6–15 shows the more common names given to the three sides of a right-angle triangle. Side C, called the **hypotenuse,** is always the longest of the three sides. Side B, called the **adjacent side,** always extends between the hypotenuse and the vertical side. An angle called **theta** (θ, a Greek letter) is formed between the hypotenuse and the adjacent side. The side that is always opposite the angle θ is called the **opposite side.**

Angles are always measured in degrees because all angles are part of a circle, like the one shown in Figure 6–16(a). A circle is divided into 360 small sections called **degrees.** In Figure 6–16(b) our right triangle has been placed within the circle. In this example the triangle occupies a 45° section of the circle and therefore theta equals 45 degrees ($\theta = 45°$). Moving from left to right in Figure 6–16(c) you will notice that angle θ increases from 5° to 85°. The length of the hypotenuse (H) in all these examples remains the same; however, the adjacent side's length decreases ($A\downarrow$) and the opposite side's length increases ($O\uparrow$) as angle θ is increased ($\theta\uparrow$). This relationship between the relative length of a triangle's sides and theta means that we do not have to know the length of two sides to calculate a third. If we have the value of just one side and the angle theta, we can calculate the length of the other two sides.

Trigonometry
The study of the properties of triangles.

Hypotenuse
The side of a right-angle triangle that is opposite the right angle.

Adjacent
The side of a right-angle triangle that has a common endpoint, in that it extends between the hypotenuse and the vertical side.

Theta
The eighth letter of the Greek alphabet, used to represent an angle.

Opposite
The side of a right-angle triangle that is opposite the angle theta.

Degree
A unit of measure for angles equal to an angle with its vertex at the center of a circle and its sides cutting off $\frac{1}{360}$ of the circumference.

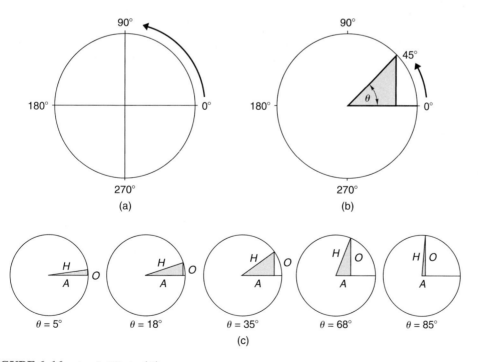

FIGURE 6–16 Angle Theta (θ).

6–3–1 Sine of Theta (Sin θ)

Sine
The trigonometric function that for an acute angle is the ratio between the leg opposite the angle when it is considered part of a right triangle, and the hypotenuse.

In the preceding section we discovered that a relationship exists between the relative length of a triangle's sides and the angle theta. **Sine** is a comparison between the length of the opposite side and the length of the hypotenuse. Expressed mathematically,

$$\text{Sine of theta } (\sin \theta) = \frac{\text{opposite side } (O)}{\text{hypotenuse } (H)}$$

Because the hypotenuse is always larger than the opposite side, the result will always be less than 1 (a decimal fraction). Let us use this formula in a few examples to see how it works.

CALCULATOR KEYS

Name: Sine key

Function: Instructs the calculator to find the sine of the displayed value (angle → value).

Example: $\sin 37° = ?$
Press keys: 3 7 SIN
Answer: 0.601815

Name: Arcsine ($\sin^{-1}$) or inverse sine sequence.

Function: Calculates the smallest angle whose sine is in the display (value → angle).

Example: invsin 0.666 = ?
Press keys: . 6 6 6 INV SIN
Answer: 41.759°

■ **TECHNICAL TRADE APPLICATION: SURVEYING**

On the set of plans shown in Figure 6–17(a), angle theta is equal to 41°, and the opposite side is equal to 20 centimeters in length. Calculate the length of the hypotenuse.

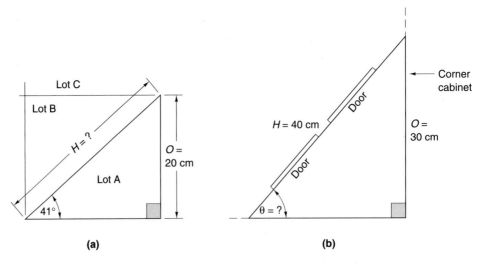

FIGURE 6–17 Sine of Theta.

■ **Solution:**

Inserting these values in our formula, we obtain the following:

$$\sin\theta = \frac{O}{H}$$

$$\sin 41° = \frac{20 \text{ cm}}{H}$$

By looking up 41° in a sine trigonometry table or by using a scientific calculator that has all the trigonometry tables stored permanently in its memory, you will find that the sine of 41° is 0.656. This value describes the fact that when $\theta = 41°$, the opposite side will be 0.656, or 65.6%, as long as the hypotenuse side. By inserting this value into our formula and transposing the formula according to the rules of algebra, we can determine the length of the hypotenuse.

$$\sin 41° = \frac{20 \text{ cm}}{H}$$

$$0.656 = \frac{20 \text{ cm}}{H}$$

Calculator sequence: [4][1][SIN]

$$0.656 \times H = \frac{20 \text{ cm} \times \cancel{H}}{\cancel{H}}$$

Multiply both sides by H.

$$\frac{\cancel{0.656} \times H}{\cancel{0.656}} = \frac{20 \text{ cm}}{0.656}$$

Divide both sides by 0.656.

$$H = \frac{20 \text{ cm}}{0.656} = 30.5 \text{ cm}$$

■ **TECHNICAL TRADE APPLICATION: DRAFTING**

Figure 6–17(b) illustrates another example; however, in this case the lengths of sides H and O are known but θ is not.

■ **Solution:**

$$\sin\theta = \frac{O}{H}$$

$$\sin\theta = \frac{30 \text{ cm}}{40 \text{ cm}}$$

$$= 0.75$$

The ratio of side O to side H is 0.75, or 75%, which means that the opposite side is 75%, or 0.75, as long as the hypotenuse. To calculate angle θ we must isolate it on one side of the equation. To achieve this we must multiply both sides of the equation by **arcsine, or inverse sine** (invsin), which does the opposite of sine.

$$\sin\theta = 0.75$$

$$\cancel{\text{invsin}}\,(\cancel{\sin}\,\theta) = \text{invsin}\, 0.75 \qquad \text{Take the inverse sine of 0.75.}$$

$$\theta = \text{invsin}\, 0.75 \qquad \text{\textit{Calculator sequence:}}\; [.][7][5][\text{INV}][\text{SIN}]$$

$$= 48.6°$$

Arcsine or Inverse Sine
The inverse function to the sine. (If y is the sine of θ, then θ is the arcsine of y.)

In summary, therefore, the sine trig functions take an angle θ and give you a number x. The inverse sine (arcsin) trig functions take a number x and give you an angle θ. In both cases, the number x is the ratio of the opposite side to the hypotenuse.

Sine: angle $\theta \rightarrow$ number x
Inverse sine: number $x \rightarrow$ angle θ

6–3–2 Cosine of Theta (Cos θ)

Cosine
The trigonometric function that for an acute angle is the ratio between the leg adjacent to the angle when it is considered part of a right triangle, and the hypotenuse.

Sine is a comparison between the opposite side and the hypotenuse, and **cosine** is a comparison between the adjacent side and the hypotenuse.

$$\text{Cosine of theta } (\cos \theta) = \frac{\text{adjacent } (A)}{\text{hypotenuse } (H)}$$

CALCULATOR KEYS

Name: Cosine key

Function: Instructs the calculator to find the cosine of the displayed value (angle → value).

Example: $\cos 26° = ?$
Press keys: [2] [6] [COS]
Answer: 0.89879

Name: Arccosine ($\cos^{-1}$) or inverse cosine sequence.

Function: Calculates the smallest angle whose cosine is in the display (value → angle).

Example: invcos 0.234 = ?
Press keys: [.] [2] [3] [4] [INV] [COS]
Answer: 76.467

■ **TECHNICAL TRADE APPLICATION: WELDING**

Figure 6–18(a) illustrates a metal support frame that is to be welded. With this right triangle the angle θ and the length of the hypotenuse are known. From this information, calculate the length of the adjacent side.

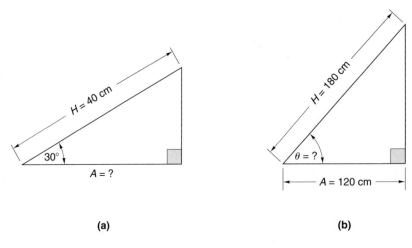

(a) (b)

FIGURE 6–18 Cosine of Theta.

■ *Solution:*

$$\cos\theta = \frac{A}{H}$$

$$\cos 30° = \frac{A}{40 \text{ cm}}$$ *Calculator sequence:* [3] [0] [cos]

$$0.866 = \frac{A}{40 \text{ cm}}$$

Looking up the cosine of 30° you will obtain the fraction 0.866. This value states that when $\theta = 30°$, the adjacent side will always be 0.866, or 86.6%, as long as the hypotenuse. By transposing this equation we can calculate the length of the adjacent side:

$$0.866 = \frac{A}{40 \text{ cm}}$$

$$40 \times 0.866 = \frac{A}{40 \text{ cm}} \times 40 \quad \text{Multiply both sides by 40.}$$

$$A = 0.866 \times 40$$
$$= 34.64 \text{ cm}$$

■ **TECHNICAL TRADE APPLICATION: CONSTRUCTION**

Calculate the angle θ in Figure 6–18(b) if $A = 120$ centimeters and $H = 180$ centimeters.

■ *Solution:*

$$\cos\theta = \frac{A}{H}$$

$$= \frac{120 \text{ cm}}{180 \text{ cm}}$$

$$= 0.667 \quad A \text{ is 66.7\% as long as } H.$$

$$\cancel{\text{invcos}}(\cancel{\cos\theta}) = \text{invcos } 0.667 \quad \text{Multiply both sides by invcos.}$$

$$\theta = \text{invcos } 0.667 \quad \text{\textit{Calculator sequence:}} \; [0][.][6][6][7][\text{INV}][\text{COS}]$$

$$= 48.2$$

The inverse cosine trig function performs the reverse operation of the cosine function:

Cosine: angle $\theta \rightarrow$ number x
Inverse cosine: number $x \rightarrow$ angle θ

6-3-3 *Tangent of Theta (Tan θ)*

Tangent is a comparison between the opposite side of a right triangle and the adjacent side.

$$\text{Tangent of theta } (\tan\theta) = \frac{\text{opposite }(O)}{\text{adjacent }(A)}$$

Tangent
The trigonometric function that for an acute angle is the ratio between the leg opposite the angle when it is considered part of a right triangle, and the leg adjacent.

CALCULATOR KEYS

Name: Tangent key

Function: Instructs the calculator to find the tangent of the displayed value (angle → value).

Example: sin 73° = ?
Press keys: [7] [3] [TAN]
Answer: 3.2709

Name: Arctangent (tan^{-1}) or inverse tangent sequence.

Function: Calculates the smallest angle whose tangent is in the display (value → angle).

Example: invtan 0.95 = ?
Press keys: [.] [9] [5] [INV] [TAN]
Answer: 43.5312

■ TECHNICAL TRADE APPLICATION: CARPENTRY

Figure 6–19(a) shows a wooden shelf support. With this right triangle θ = 65° and the opposite side is 43 centimeters. Calculate the length of the adjacent side.

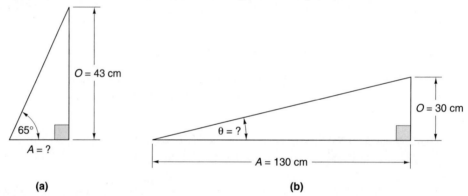

(a)   (b)

FIGURE 6–19 Tangent of Theta.

■ *Solution:*

$$\tan \theta = \frac{O}{A}$$

$$\tan 65° = \frac{43 \text{ cm}}{A}$$ *Calculator sequence:* [6] [5] [TAN]

$$2.14 = \frac{43 \text{ cm}}{A}$$ Whenever θ = 65°, the opposite side will be 2.14 times longer than the adjacent side.

$$2.14 \times A = \frac{43 \text{ cm}}{A} \times A$$ Multiply both sides by *A*.

$$\frac{\cancel{2.14} \times A}{\cancel{2.14}} = \frac{43 \text{ cm}}{2.14}$$ Divide both sides by 2.14.

$$A = \frac{43 \text{ cm}}{2.14}$$

$$= 20.1 \text{ cm}$$

■ **TECHNICAL TRADE APPLICATION: CONSTRUCTION**

Calculate the angle θ in Figure 6–19(b) for the concrete handicapped ramp if $O = 30$ centimeters and $A = 130$ centimeters.

■ *Solution:*

$$\tan \theta = \frac{O}{A}$$

$$= \frac{30 \text{ cm}}{130 \text{ cm}}$$

$$= 0.231 \qquad O \text{ is 0.231 or 23.1\% as long as } A.$$

$$\text{invtan}(\tan \theta) = \text{invtan } 0.231 \qquad \text{Multiply both sides by invtan.}$$

$$\theta = \text{invtan } 0.231 \qquad \textit{Calculator sequence:} \boxed{0}\boxed{.}\boxed{2}\boxed{3}\boxed{1}\boxed{\text{INV}}\boxed{\text{TAN}}$$

$$= 12.99 \text{ or } 13°$$

6–3–4 Summary

As you have seen, trigonometry involves the study of the relationships among the three sides (O, H, A) of a right triangle and also the relationships among the sides of the right triangle and the number of degrees contained in the angle theta (θ).

If the lengths of two sides of a right triangle are known, and the length of the third side is needed, remember that

$$H^2 = O^2 + A^2$$

If the angle θ is known along with the length of one side, or if angle θ is needed and the lengths of the two sides are known, one of the three formulas can be chosen based on which variables are known and what is needed.

$$\sin \theta = \frac{O}{H} \qquad \cos \theta = \frac{A}{H} \qquad \tan \theta = \frac{O}{A}$$

When I was introduced to trigonometry, my mathematics professor spent 15 minutes having the whole class practice what he described as an old Asian war cry that went like this: "SOH CAH TOA." After he explained that it wasn't a war cry but in fact a memory aid to help us remember that SOH was in fact $\sin \theta = O/H$, CAH was $\cos \theta = A/H$, and TOA was $\tan \theta = O/A$, we understood the method in his madness.

■ **TECHNICAL TRADE APPLICATION: ELECTRICAL**

The **sine wave** is the most common type of electrical waveform. It is the natural output of a generator that converts a mechanical input, in the form of a rotating shaft, into an electrical output in the form of a sine wave. In fact, for one cycle of the input shaft, the generator will produce one sinusoidal AC voltage waveform, as shown in Figure 6–20. When the input shaft of the generator is at 0°, the AC output is 0 volts. As the shaft is rotated through 360° the AC output voltage will rise to a maximum or peak positive voltage at 90°, fall back to 0 volts at 180°, then reach a maximum or peak negative voltage at 270°, and finally return to 0 volts at 360°. If this AC voltage is applied across a closed circuit, it produces a current that continually reverses or alternates in direction. (It switches from being a positive voltage to a negative voltage, back to a positive voltage, and so on.) The sine wave is the most common type of waveform shape, and why the name *sine* was given to this wave needs to be explained further.

Sine Wave

A waveform that represents periodic oscillations in which the amplitude of displacement at each point is proportional to the sine of the phase angle of the displacement and that is visualized as a sine curve.

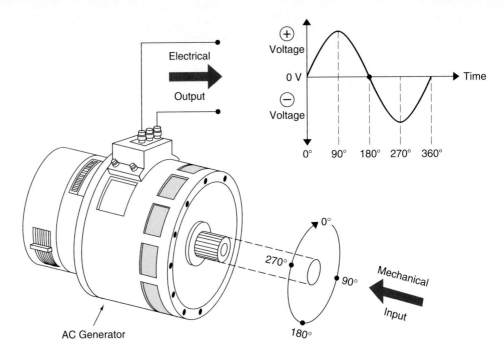

FIGURE 6–20 Degrees of a Sine Wave.

Figure 6–21(a) shows the correlation between the 360° of a circle and the 360° of a sine wave. Within the circle a triangle has been drawn to represent the right-angle triangle formed at 30°. The hypotenuse side will always remain the same length throughout 360° and will equal the maximum voltage or peak of the sine wave. The opposite side of the triangle is equal to the amplitude vector of the sine wave at 30°. To calculate the amplitude of the opposite side, and therefore the amplitude of the sine wave at 30°, we use the sine of theta formula discussed previously:

$$\sin \theta = \frac{\text{opposite}}{\text{hypotenuse}}$$

$$\sin 30° = \frac{O}{H} \qquad \textit{Calculator sequence:} \boxed{3}\boxed{0}\boxed{\text{SIN}}$$

$$0.500 = \frac{O}{H}$$

This equation tells us that at 30°, the opposite side is 0.500 or half the size of the hypotenuse. At 30°, therefore, the amplitude or magnitude of the sine wave will be 0.5 (50%) of the peak value.

Figure 6–21(b) lists the sine values at 15° increments. Figure 6–21(c) shows an example of a 10 volt peak sine wave. At 15°, a sine wave will always be at 0.259 (sine of 15°) of the peak value, which for a 10 volt sine wave will be 2.59 volts (0.259 × 10 volts = 2.59 volts). At 30°, a sine wave will have increased to 0.500 (sine of 30°) of the peak value. At 45°, a sine wave will be at 0.707 of the peak, and so on. The sine wave is called a sine wave because it changes in amplitude at the same rate as the sine trigonometric function.

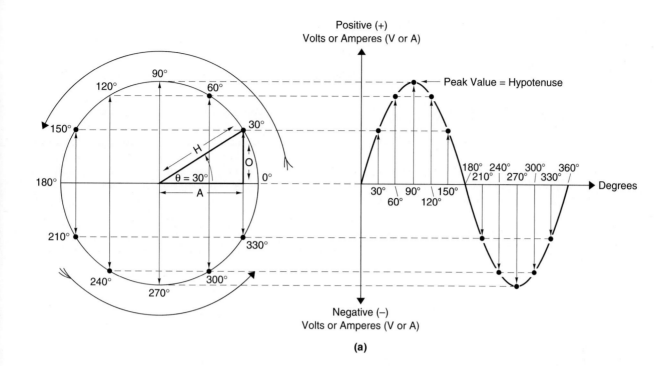

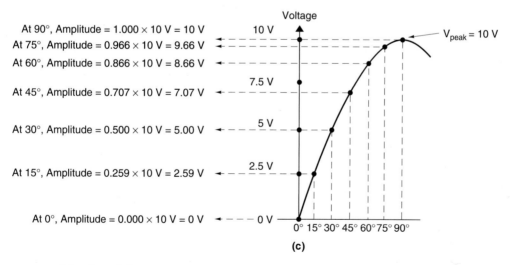

FIGURE 6–21 Meaning of Sine.

SELF-TEST EVALUATION POINT FOR SECTION 6–3

Now that you have completed this section, you should be able to:

- **Objective 9.** Define the trigonometric terms opposite, adjacent, hypotenuse, theta, sine, cosine, and tangent.

- **Objective 10.** Demonstrate how the sine, cosine, and tangent trigonometric functions can be used to calculate (1) the length of an unknown side of a right-angle triangle if the length of another side and the angle theta are known or (2) the angle theta of a right-angle triangle if the lengths of two sides of the triangle are known.

Use the following questions to test your understanding of Section 6–3.

1. Calculate the length of the unknown side in the following triangles.

   a.
   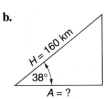

   b.
   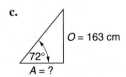

   c.
   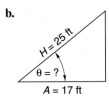

2. Calculate the angle θ for the following right-angle triangles.

   a.
   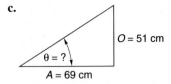

   b.

   c.

Use the following technical trade questions to test your understanding of practical applications of Section 6–3.

1. **Electronics** Some components in electronics circuits, such as resistors, consume power from a source and dissipate it in the form of heat. This power consumption is known as *resistive power* or *true power*. Other components, like capacitors, take power from the source at one time and then return it to the circuit at another time. This power consumption is known as *reactive power* or *imaginary power*.

    When capacitors and resistors are together in a circuit, they consume power at different times, and so you cannot simply add together the resistive and reactive power to find the total. The resulting apparent power is obtained by vectorially adding the resistive and reactive power.

    Referring to the right triangle shown, determine:
    a. P_A when $P_R = 12$ watts and $P_X = 14$ watts.
    b. Angle θ, using sine.
    c. Since all three side lengths are known, show how the angle can be determined using cosine and tangent.

2. **Manufacturing** In the automated manufacturing process shown, a connecting conveyor is needed to transport parts from one system to another. Determine the length of the conveyor needed.

    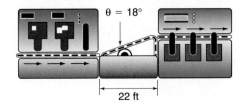

3. **Machining** Referring to the part shown, determine:
    a. the length of the taper (H).
    b. Angle θ.

    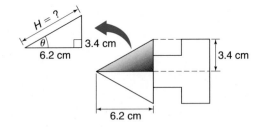

194   CHAPTER 6 / GEOMETRY AND TRIGONOMETRY

6–4 SOLID FIGURES

As mentioned previously, a plane figure is a flat surface that has only two dimensions—length and width. A **solid figure**, on the other hand, is a three-dimensional figure having length, width, and height.

Solid Figure
A 3-dimensional figure having length, width, and height.

6–4–1 *Prisms*

Prism
A solid figure made up of plane figure faces; having at least one pair of parallel surfaces.

A **prism** is a solid figure made up of plane figure faces and has at least one pair of parallel surfaces. Figure 6–22 details the facts and formulas for the prism. The visual breakdown in

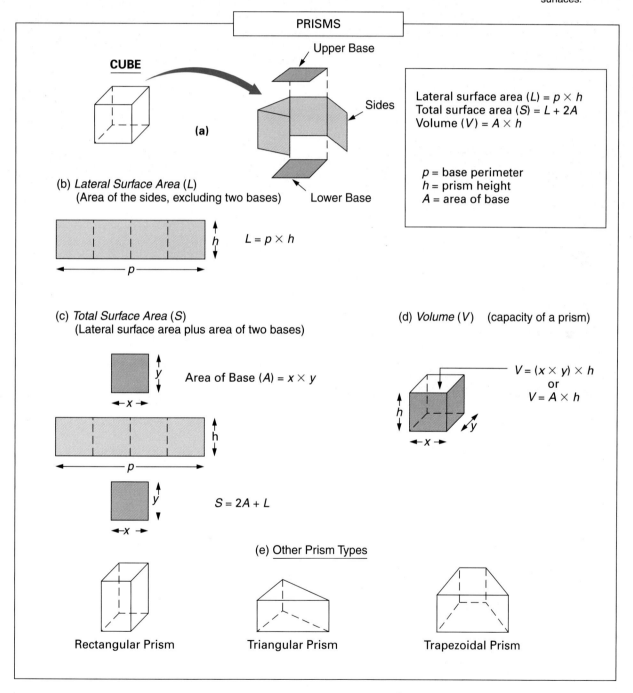

FIGURE 6–22 Prisms.

Lateral Surface Area
The side area of a solid figure.

Total Surface Area
The lateral surface area of a solid figure plus the area of its two bases.

Volume
The space within a solid figure, or its holding capacity.

Cylinder
A solid figure with curved side walls extending between two identical circular bases.

Sphere
A globular solid figure in which all points on the surface are at equal distance from the center.

Figure 6–22(a) shows how a cube is made up of an upper base, lower base, and a set of sides. The **lateral surface area (L)** of the cube is the area of the sides only, as shown in Figure 6–22(b). The **total surface area (S)** is the lateral surface area of the cube plus the area of the two bases, as shown in Figure 6–22(c). The holding capacity or **volume** of the cube measures the space inside the prism and is described visually in Figure 6–22(d). Figure 6–22(e) shows examples of other prism types.

6–4–2 *Cylinders*

A **cylinder** is a solid figure with curved side walls extending between two identical circular bases. Figure 6–23 details the facts and formulas for the cylinder. As with prisms, plane figure formulas are used to calculate the cylinder's lateral surface area and base area, and as expected, the volume of a cylinder is the product of its base area and height.

6–4–3 *Spheres*

A **sphere** is a globular solid figure in which all points on the surface are at an equal distance from the center. Figure 6–24 lists the formulas for a sphere along with a visual breakdown for a sphere's surface area and volume.

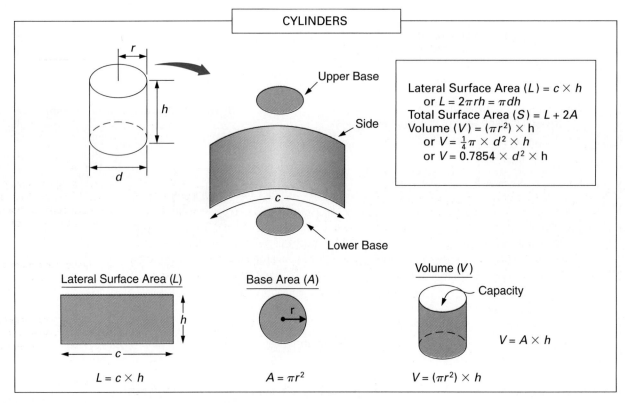

FIGURE 6–23 Cylinders.

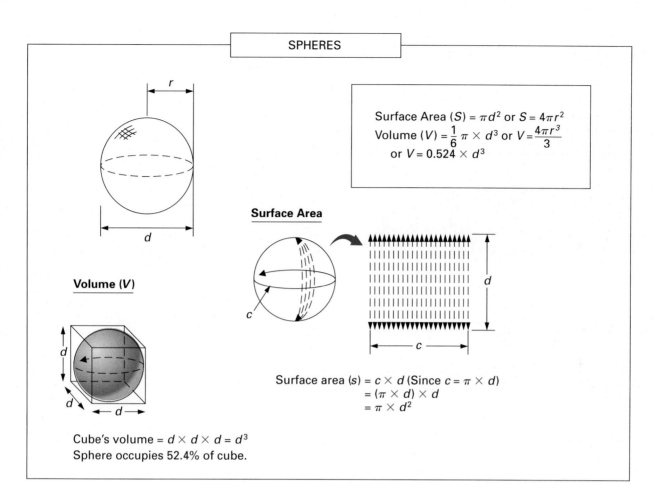

FIGURE 6–24 Spheres.

6–4–4 Pyramids and Cones

Figure 6–25(a) lists the formulas for a **pyramid,** which has only one base and three or more lateral surfaces that taper up to a single point called an **apex.** Figure 6–25(a) lists the formulas for a **cone,** which has a circular base and smooth surface that extends up to an apex. Figures 6–25(b) and 6–26(b) list the formulas for the frustum of a pyramid and the frustum of a cone. A **frustum** is the base section of a solid pyramid or cone and is formed by cutting off the top of the solid figure so that the upper base and lower base are parallel to one another.

Pyramid

A solid figure having only one base and 3 or more lateral surfaces that taper up to a single point.

Apex

The uppermost point or tip of a solid figure.

Cone

A solid figure having a circular base and smooth surface that extends up to an apex.

Frustum

The base section of a solid pyramid or cone.

PYRAMIDS AND FRUSTUMS OF PYRAMIDS

(a) PYRAMID

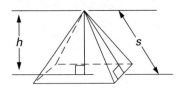

Lateral Surface Area $(L) = \frac{1}{2} ps$
Volume $(V) = \frac{1}{3} Ah$

p = perimeter of base s = slant height
A = area of base h = height

Lateral Surface Area (L)

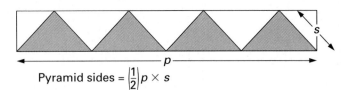

Pyramid sides = $\left(\frac{1}{2}\right) p \times s$

Volume (V)

Cube Volume = $A \times h$.
Pyramid occupies $\frac{1}{3}$ of cube.

(b) FRUSTUM OF PYRAMID

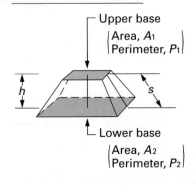

- Upper base (Area, A_1; Perimeter, P_1)
- Lower base (Area, A_2; Perimeter, P_2)

Lateral surface area $(L) = \frac{1}{2}(P_1 + P_2)s$
Volume $(V) = \frac{1}{3} h(A_1 + A_2 + \sqrt{A_1 \times A_2})$

FIGURE 6–25 Pyramids and Frustums of Pyramids.

CONES AND FRUSTUMS OF CONES

(a) CONE

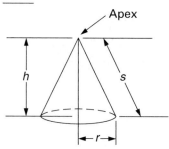

Lateral Surface Area $(L) = \pi rs$ or $L = \frac{1}{2}\pi ds$
Volume $(V) = \frac{1}{3}\pi r^2 h$ or $V = \frac{1}{12}\pi d^2 h$

r = base radius h = height
s = slant height d = base diameter

(b) FRUSTUM OF CONE

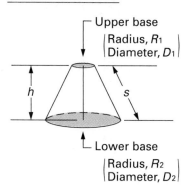

Lateral Surface Area $(L) = \pi s (r_1 + r_2)$
or $L = \frac{1}{2}\pi s (d_1 + d_2)$

Volume $(V) = \frac{1}{3}\pi h (r_1^2 + r_2^2 + (r_1 r_2))$
or $V = \frac{1}{12}\pi h (d_1^2 + d_2^2 + (d_1 d_2))$

FIGURE 6–26 Cones and Frustums of Cones.

SELF-TEST EVALUATION POINT FOR SECTION 6–4

Now that you have completed this section, you should be able to:

■ **Objective 11.** *Define the term* solid figure, *and identify the different types.*

■ **Objective 12.** *List the formulas for determining the lateral surface area, total surface area, and volume of solid figures.*

Use the following questions to test your understanding of Section 6–4:

1. What is the difference between a plane figure and a solid figure?
2. What is the difference between lateral surface area and total surface area?
3. Sketch the following solid figures and list the formulas for lateral surface area, total surface area and volume.
   a. Prism
   b. Cylinder
   c. Sphere
   d. Pyramid
   e. Cone
4. What is a frustum?

Use the following technical trade questions to test your understanding of practical applications of Section 6–4.

1. **Manufacturing** Determine the weight of the steel rectangular prism if it has a density of 0.28 pounds per cubic inch (weight = density × volume).

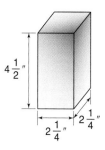

SECTION 6–4 / SOLID FIGURES   **199**

2. **Painting** If 1 gallon of paint covers 300 square feet, how many gallons would be needed to paint the office building shown?

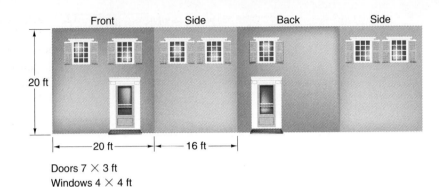

Doors 7 × 3 ft
Windows 4 × 4 ft

3. **Landscaping** To prepare for topsoil and sod, a circular section measuring 35 feet in diameter will have $1\frac{1}{2}$ feet of dirt removed. How many cubic feet is this?

4. **Roofing** A roof's base measures 40 meters square and has a slant height of 24 meters. If roofing materials cost $5.75 per square meter, how much will it cost to roof the building?

5. **Machining** If aluminum has a density of 0.093 pound per cubic inch, what is the weight of the machined capsule shown?

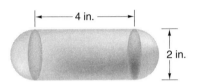

REVIEW QUESTIONS

Multiple Choice Questions

1. What instrument can be used to measure an angle?
   - **a.** Compass
   - **b.** Protractor
   - **c.** Ruler
   - **d.** None of the above

2. A right angle is equal to:
   - **a.** 45°
   - **b.** 100°
   - **c.** 180°
   - **d.** 90°

3. A protractor can be used to:
   - **a.** Measure an angle
   - **b.** Draw an angle
   - **c.** Both (a) and (b)
   - **d.** None of the above

4. A _____ is a quadrilateral parallelogram, that has four sides that are all of equal length.
   - **a.** Square
   - **b.** Rectangle
   - **c.** Trapezoid
   - **d.** Prism

5. Trigonometry is the study of:
   - **a.** Triangles
   - **b.** Angles
   - **c.** Trigonometric functions
   - **d.** All the above

6. The Pythagorean theorem states that the _____ of the length of the hypotenuse of a right-angle triangle equals the sum of the _____ of the lengths of the other two sides.
   - **a.** Square, square roots
   - **b.** Square root, squares
   - **c.** Square, squares
   - **d.** Square root, square roots

7. A vector is an arrow whose length is used to represent the _____ of a quantity and whose point is used to indicate the same quantity's _____.
   - **a.** Magnitude, size
   - **b.** Magnitude, direction
   - **c.** Direction, size
   - **d.** Direction, magnitude

8. If two vectors in a vector diagram are not working together, or are out of phase with each other, they must be _____ to obtain a _____ vector.
   a. Added, resultant
   b. Multiplied by one another, parallel
   c. Vectorially added, resultant
   d. Both (a) and (b)

9. The _____ of a right-angle triangle is always the longest side.
   a. Opposite side
   b. Adjacent side
   c. Hypotenuse
   d. Angle theta

10. The _____ of a right-angle triangle is always across from the angle theta.
    a. Opposite side
    b. Adjacent side
    c. Hypotenuse
    d. Angle theta

11. The _____ of a right-angle triangle is always formed between the hypotenuse and the adjacent side.
    a. Opposite side   c. Hypotenuse
    b. Adjacent side   d. Angle theta

12. Which trigonometric function can be used to determine the length of the hypotenuse if the angle theta and the length of the opposite side are known?
    a. Tangent
    b. Sine
    c. Cosine

13. Which trigonometric function can be used to determine the angle theta when the lengths of the opposite and adjacent sides are known?
    a. Tangent
    b. Sine
    c. Cosine

14. Which trigonometric function can be used to determine the length of the adjacent side when the angle theta and the length of the hypotenuse are known?
    a. Tangent
    b. Sine
    c. Cosine

15. A plane figure has _____ dimensions, while a solid figure has _____ dimensions.
    a. 3, 4
    b. 2, 4
    c. 2, 3
    d. 3, 2

Communication Skill Questions

16. Define the following terms: (6–1 to 6–4)
    a. Geometry   c. Plane figure
    b. Trigonometry   d. Solid figure

17. Describe how to use a protractor to draw and measure an angle. (6–1)

18. Sketch the following plane figures and list their associated formulas for calculating perimeter and area: (6–2)
    a. Square   f. Right triangle
    b. Rectangle   g. Scalene triangle
    c. Rhombus   h. Equilateral triangle
    d. Parallelogram   i. Isosceles triangle
    e. Trapezoid   j. Circle

19. What are the number of degrees in each vertex of a: (6–2–5)
    a. Square   d. Hexagon
    b. Equilateral triangle   e. Octagon
    c. Pentagon

20. Describe the purpose of Pythagoras's theorem. (6–2–2)

21. Demonstrate how the sine, cosine and tangent trigonometric functions can be used to calculate: (6–3)
    a. The length of an unknown side of a right triangle when the length of another side is known and the angle theta is known
    b. The angle theta of a right triangle if the lengths of two sides of the triangle are known

22. Sketch the following solid figures and list their associated formulas: (6–4)
    a. Prism   d. Pyramid
    b. Cylinder   e. Cone
    c. Sphere

23. Why is it necessary to have inverse trigonometric functions? (6–3)

24. What is a sine wave? (6–3)

25. What is a frustum? (6–4)

Practice Problems

26. Sketch the following:
    a. A line segment   e. Two perpendicular lines
    b. Two parallel lines   f. An angle of 45°
    c. A right angle   g. A vertical line
    d. An acute angle   h. An obtuse angle

27. Calculate the length of the hypotenuse (C side) of the following right triangles.
    a. $A = 20$ mi, $B = 53$ mi
    b. $A = 2$ km, $B = 3$ km
    c. $A = 4$ in., $B = 3$ in.
    d. $A = 12$ mm, $B = 12$ mm

28. Calculate the length or magnitude of the resultant vectors in the following vector diagrams.

a.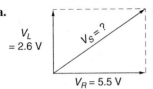

b.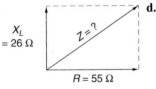

c.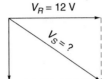

d.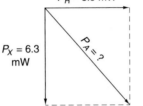

29. Calculate the value of the following trigonometric functions.

a. sin 0°
b. sin 30°
c. sin 45°
d. sin 60°
e. sin 90°
f. cos 0°
g. cos 30°
h. cos 45°
i. cos 60°
j. cos 90°
k. tan 0°
l. tan 30°
m. tan 45°
n. tan 60°
o. tan 90°

30. Calculate angle θ from the function given.

a. $\sin \theta = 0.707, \theta = ?$
b. $\sin \theta = 0.233, \theta = ?$
c. $\cos \theta = 0.707, \theta = ?$
d. $\cos \theta = 0.839, \theta = ?$
e. $\tan \theta = 1.25, \theta = ?$
f. $\tan \theta = 0.866, \theta = ?$

31. Name each of the triangle's sides, and calculate the unknown values.

a.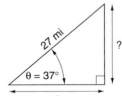

b.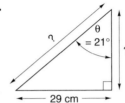

c.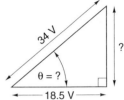

32. If a cylinder has a radius of 3 centimeters and a height of 10 centimeters, calculate its:

a. Lateral surface area
b. Total surface area
c. Volume

33. If a sphere has a diameter of 2 inches, what would be its surface area and volume?

34. What is the lateral surface area of a rectangular prism that has a base perimeter of 40 centimeters and a height of 12 centimeters?

35. Calculate the lateral surface area of a pyramid that has a base perimeter of 24 inches and a slant height of 8 inches.

Math Application Practice Problems

36. **Landscaping** Determine the area of the grass sod strips shown.

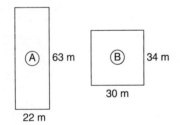

37. **Aerospace** Determine the banking angle b of the aircraft shown, and the angle a. What is the sum of angles a and b?

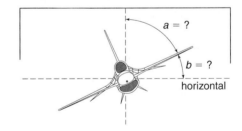

38. **Landscaping** Determine angle x for the lawnmower shown.

39. **Construction** The *plummet* was used by the Egyptians to tell whether stones were level. Describe how this could be used.

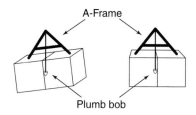

40. **Automotive** What triangles are formed within the automobile jack shown?

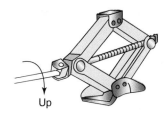

41. **Electronics** The size of a television screen is measured diagonally as shown. From the measurements given, determine the length of side x.

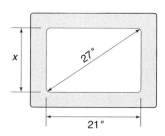

42. **Carpentry** From the measurements given, determine the unknown length (x) of the roof frame.

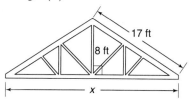

43. **Automotive** The drum shown is used to store waste oil. What is its volume?

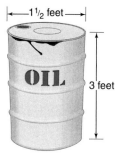

44. **Plumbing** How many cubic feet of volume does the water heater shown have?

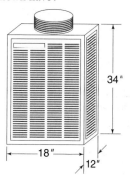

45. **Automotive** An engine's *compression ratio* is the ratio of a cylinder's bottom-dead-center (BDC) volume to its top-dead-center (TDC) volume. What is the compression ratio of the engine cylinder shown?

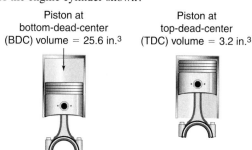

Web Site Questions

Go to the Web site http://www.prenhall.com/cook, select the textbook *Mathematics for the Technical Trades*, select this chapter, and then follow the instructions when answering the multiple-choice practice problems.

Answers to Self-Test Evaluation Points

APPENDIX A

STEP 1-1
1. 10
2. $3 \times 10,000 = 30,000$
   $5 \times 1,000 = 5,000$
   $7 \times 100 = 700$
   $2 \times 10 = 20$
   $9 \times 1 = 9$
   $35,729$
3. See Section 1-1-3.
4. 499,997
499,998	1
499,999	2
500,000	3
500,001	4
500,002	5

   The units, tens, hundreds, thousands, and ten-thousands columns will reset and carry when the count is advanced by 3. The final count will be 500,002.

STEP 1-1 Technical Trade Questions
1. Count up 9 from 147 = 156 lb
2. (a) x to y = 24 cm
   (b) y to z = 36 cm
3. Front = 4 cm
   Rear = 2 cm
4. 20 cm − 7 cm = 13 cm
   13 cm + 2 cm = 15 cm
5. 7 plus 7, plus 7 = 21
6. $107 plus $8 = $115

STEP 1-2
1. (a) $\frac{6}{8}$ (b) $\frac{4}{8}$ (c) $\frac{5}{16}$
2. (a) $\frac{2}{8}$ (b) $\frac{4}{8}$ (c) $\frac{11}{16}$

STEP 1-2 Technical Trade Questions
1. (d)
2. (d)
3. Pitch = $\frac{\text{Rise}}{\text{Run}} = \frac{8}{20}$
4. (a) $2\frac{1}{2}$ amps
   (b) $4\frac{1}{4}$ amps
   (c) $5\frac{3}{4}$ amps
   (d) $6\frac{7}{8}$ amps
   (e) $8\frac{3}{8}$ amps
5. 30 rolls − 14 = 16 rolls remaining
   $\frac{16}{30}$ (16 rolls of 30 remaining)
6. (c) $\frac{7}{8}$ in
7. 4 oz. of fertilizer + 68 oz. of water = 72 ounces
   $\frac{4}{72}$

8. They are the same thickness—$\frac{2}{4} = \frac{1}{2}$.
9. $\frac{4}{8} = \frac{1}{2}$, therefore, 1 unit every 2 hours
10. $5 \times \$3.97 = \$19.85$
    $\frac{\text{Paper Cost}}{\text{Cartridge Cost}} = \frac{\$19.85}{\$89.95}$

STEP 1-3
1. $\frac{4}{6} + \frac{1}{6} = \frac{4+1}{6} = \frac{5}{6}$
2. $\frac{3}{64} + \frac{2}{64} + \frac{25}{64} = \frac{3+2+25}{64} = \frac{30 \div 2}{64 \div 2} = \frac{15}{32}$
3. $\frac{3}{9} + \frac{6}{18} = \frac{6+6}{18} = \frac{12 \div 6}{18 \div 6} = \frac{2}{3}$
4. $\frac{1}{3} + \frac{1}{4} + \frac{3}{15} = \frac{?}{?}$

2	3	4	15
2	3	2	15
3	3	1	15
5	1	1	5
	1	1	1

   $\rightarrow 2 \times 2 \times 3 \times 5 = 60$

   $\frac{1}{3} + \frac{1}{4} + \frac{3}{15} = \frac{20 + 15 + 12}{60} = \frac{47}{60}$
5. $\frac{3}{4} + \frac{4}{5} = \frac{15 + 16}{20} = \frac{31}{20} = 1\frac{11}{20}$

2	4	5
2	2	5
5	1	5
	1	1

   $\rightarrow 2 \times 2 \times 5 = 20$

6. $\frac{9}{12} + \frac{4}{24} + \frac{3}{4} = \frac{18+4+18}{24} = \frac{40}{24} = 1\frac{16 \div 8}{24 \div 8} = 1\frac{2}{3}$
7. $\frac{7}{9} + \frac{5}{9} + \frac{4}{18} = \frac{14+10+4}{18} = \frac{28}{18} = 1\frac{10 \div 2}{18 \div 2} = 1\frac{5}{9}$
8. $\frac{15}{6} = 2\frac{3}{6} \div \frac{3}{3} = 2\frac{1}{2}$
9. $\frac{4}{24} = \frac{1}{6}$
10. $\frac{25}{100} \div \frac{25}{25} = \frac{1}{4}$

STEP 1-3 Technical Trade Questions
1. $\frac{3}{4} + 1\frac{1}{2} = 2\frac{1}{4}$
2. $2\frac{1}{2} + 4\frac{1}{3} + \frac{1}{4} + 3\frac{1}{6} = 10\frac{1}{4}$
3. $\frac{1}{8} + \frac{7}{16} + \frac{3}{8} = \frac{15}{16}$
4. $9\frac{3}{4} + 8\frac{1}{3} + 16\frac{1}{8} + 8\frac{1}{2} = 42\frac{17}{24}$
5. $14\frac{1}{4} + 7\frac{1}{3} + 6\frac{1}{8} = 27\frac{17}{24}$
6. $\frac{3}{32} + \frac{1}{4} + \frac{3}{32} + \frac{2}{3} = \frac{106}{96} = \frac{53}{48} = 1\frac{5}{48}$

205

STEP 1-4

1. $\dfrac{4}{32} - \dfrac{2}{32} = \dfrac{4-2}{32} = \dfrac{2 \div 2}{32 \div 2} = \dfrac{1}{16}$

2. $\dfrac{4}{5} - \dfrac{3}{10} = \dfrac{8-3}{10} = \dfrac{5 \div 5}{10 \div 5} = \dfrac{1}{2}$

3. $\dfrac{6}{12} - \dfrac{4}{16} = \dfrac{24-12}{48} = \dfrac{12}{48} = \dfrac{1}{4}$

$$\begin{array}{c|cc}
2 & 12 & 16 \\
2 & 6 & 8 \\
2 & 3 & 4 \\
2 & 3 & 2 \\
3 & 3 & 1 \\
 & 1 & 1
\end{array}$$

$\rightarrow 2 \times 2 \times 2 \times 2 \times 3 = 48$

4. $15\dfrac{1}{3} - 2\dfrac{4}{12} = 13\dfrac{4-4}{12} = 13\dfrac{0}{12} = 13$

5. $15\dfrac{15}{36} - 2\dfrac{31}{36} = 3\dfrac{15-31}{36}$ ← need to borrow

$= 4\left(\dfrac{36}{36}\right) + \dfrac{15}{36} - 2\dfrac{31}{36}$

$= 4\dfrac{51}{36} - 2\dfrac{31}{36} = 2\dfrac{51-31}{36} = 2\dfrac{20}{26} = 2\dfrac{5}{9}$

STEP 1-4 Technical Trade Questions

1. $4\dfrac{1}{8} - 3\dfrac{3}{4} = \dfrac{3}{8}$

2. $8\dfrac{3}{4} - 6\dfrac{1}{3} = 2\dfrac{5}{12}$

3. $3\dfrac{5}{16} - 2\dfrac{7}{32} = 1\dfrac{3}{32}$

4. $2\dfrac{1}{2} - \dfrac{1}{4} = 2\dfrac{1}{4}$ watts

5. $\dfrac{750}{1000} - \dfrac{14}{1000} = \dfrac{736}{1000}$

6. $\dfrac{12}{16} - \dfrac{5}{8} = \dfrac{2}{16}$

STEP 1-5

1. $\dfrac{1}{3} \times \dfrac{2}{3} = \dfrac{1 \times 2}{3 \times 3} = \dfrac{2}{9}$

2. $\dfrac{4}{8} \times \dfrac{2}{9} = \dfrac{8 \div 8}{72 \div 8} = \dfrac{1}{9}$

3. $\dfrac{4}{17} \times \dfrac{3}{34} = \dfrac{12}{578} = \dfrac{6}{289}$

4. $\dfrac{1}{16} \times 5\dfrac{3}{7} = \dfrac{1}{6} \times \dfrac{38}{7} = \dfrac{38}{112} = \dfrac{19}{56}$

5. $2\dfrac{1}{7} \times 7\dfrac{2}{3} = \dfrac{15}{7} \times \dfrac{23}{3} = \dfrac{345}{21} = 16\dfrac{9}{21} = 16\dfrac{3}{7}$

STEP 1-5 Technical Trade Questions

1. $14 \times 2\dfrac{1}{2} = 35''$

   $13 \times \dfrac{3}{8} = 4\dfrac{7}{8}''$

   wall height $= 39\dfrac{7}{8}''$

2. $34 \times \dfrac{7}{12} = 19\dfrac{5}{6}''$

3. (a) $\dfrac{1}{4} \times 360° = 90°$

   (b) $\dfrac{1}{3} \times 360° = 120°$

   (c) $\dfrac{1}{5} \times 360° = 72°$

4. $5 \times \dfrac{1}{18}'' = \dfrac{5}{18}''$

5. $2\dfrac{1}{4} \times 32\dfrac{2}{5} = 72\dfrac{9}{10}$

6. $7\dfrac{3}{4} \times \dfrac{2}{3} = 5\dfrac{1}{6}$

STEP 1-6

1. $\dfrac{2}{3} \div \dfrac{1}{12} = \dfrac{2}{3} \times \dfrac{12}{1} = \dfrac{24}{3} = 8$

2. $3\dfrac{1}{4} \div \dfrac{3}{4} = \dfrac{13}{4} \div \dfrac{3}{4} = \dfrac{13}{4} \times \dfrac{4}{3} = \dfrac{52}{12} = 4\dfrac{4}{12} = 4\dfrac{1}{3}$

3. $8\dfrac{16}{20} \div 4\dfrac{4}{20} = \dfrac{176}{20} \div \dfrac{84}{20} = \dfrac{176}{20} \times \dfrac{20}{84} = \dfrac{3520}{1680}$

   $= 2\dfrac{160 \div 10}{1680 \div 10} = 2\dfrac{16 \div 4}{168 \div 4} = 2\dfrac{4 \div 2}{42 \div 2} = 2\dfrac{2}{21}$

4. $4\dfrac{3}{8} \div 2\dfrac{5}{16} = \dfrac{35}{8} \div \dfrac{37}{16} = \dfrac{35}{8} \times \dfrac{16}{37}$

   $= \dfrac{560}{296} = 1\dfrac{264 \div 4}{296 \div 4} = 1\dfrac{66 \div 2}{74 \div 2} = 1\dfrac{33}{37}$

STEP 1-6 Technical Trade Questions

1. $16 \div 2\dfrac{3}{4} = 5\dfrac{9}{11}$

2. $400 \div 34\dfrac{1}{3} = 11\dfrac{67}{103}$

3. $2\dfrac{1}{2} \div \dfrac{1}{16} = 40$

4. $25 \div 2\dfrac{1}{4} = 11\dfrac{1}{9}$

5. $28\dfrac{1}{2} \div 8\dfrac{1}{3} = 3\dfrac{21}{50}$

   $3 \times 8 = 24$

6. $16\dfrac{1}{4} \div 2 = 8\dfrac{1}{8}''$

STEP 1-7

1. $\dfrac{16}{\cancel{27}} \times \dfrac{\cancel{27}}{240} = \dfrac{\cancel{16}}{1} \times \dfrac{1}{\cancel{240}} = \dfrac{1}{1} \times \dfrac{1}{15} = \dfrac{1}{15}$

2. $\dfrac{8}{\cancel{17}} \times \dfrac{3}{\cancel{4}} = \dfrac{2}{17} \times \dfrac{3}{1} = \dfrac{6}{17}$

3. $\dfrac{15}{32} + \dfrac{30}{4} = \dfrac{\cancel{15}}{32} \times \dfrac{4}{\cancel{30}} = \dfrac{1}{\cancel{32}} \times \dfrac{\cancel{4}}{2}$

   $= \dfrac{1}{8} \times \dfrac{1}{2} = \dfrac{1}{16}$

4. $6\dfrac{2}{3} \times 2\dfrac{7}{10} = \dfrac{\cancel{20}}{3} \times \dfrac{27}{\cancel{10}} = \dfrac{2}{\cancel{3}} \times \dfrac{\cancel{27}}{1}$

   $= \dfrac{2}{1} \times \dfrac{9}{1} = \dfrac{18}{1} = 18$

5. $6\dfrac{1}{2} \div \dfrac{9}{32} = \dfrac{13}{\cancel{2}} \times \dfrac{\cancel{32}}{9} = \dfrac{13}{1} \times \dfrac{16}{9} = \dfrac{208}{9} = 23\dfrac{1}{9}$

STEP 1-7 Technical Trade Questions

1. Pitch = $\dfrac{\text{Rise}}{\text{Run}} = \dfrac{12'}{36'} \dfrac{\div\ 3}{\div\ 3} = \dfrac{4}{12}$ Pitch = 4

2. $12\dfrac{6}{8} = 12\dfrac{3}{4}$

3. $16\dfrac{3}{8} + 4\dfrac{1}{8} = 20\dfrac{4}{8} = 20\dfrac{1}{2}$

4. $8\dfrac{1}{2} + 40\dfrac{3}{8} + 24\dfrac{1}{4} + 18\dfrac{6}{16} + 16\dfrac{3}{4} = 108\dfrac{4}{16} = 108\dfrac{1}{4}$

5. $4\dfrac{3}{32} - 3\dfrac{27}{32} = \dfrac{131}{32} - \dfrac{123}{32} = \dfrac{8}{32} = \dfrac{8 \div 8}{32 \div 8} = \dfrac{1}{4}$

6. $9\dfrac{7}{8} \times 4 = \dfrac{79}{8} \times 4 = \dfrac{316}{8} = 39\dfrac{4}{8} = 39\dfrac{1}{2}$

STEP 2-1

1. $1 \times 100 = 100$
    $7 \times 10 = 70$
    $8 \times 1 = 8$
    $6 \times 0.1 = 0.6$
    $4 \times 0.01 = 0.04$
    $9 \times 0.001 = 0.009$
    $\overline{178.649}$

2. (a) $1\dfrac{5}{2} = 1 + (5 \div 2) = 1 + 2.5 = 3.5$

   (b) $192\dfrac{3}{4} = 192 + (3 \div 4) = 192 + 0.75 = 192.75$

   (c) $67\dfrac{6}{9} = 67 + (6 \div 9) = 67 + 0.667 = 67.667$

3. (a) $\dfrac{1}{2500} = 0.0004$

   (b) $\dfrac{1}{0.25} = 4$

   (c) $\dfrac{1}{0.000025} = 40{,}000$

4. (a) $7.25 = 7\dfrac{4}{16} \left(\dfrac{1}{16} = 0.0625 \right)$

   (b) $156.90625 = 156\dfrac{29}{32} \left(\dfrac{1}{32} = 0.03125 \right)$

STEP 2-1 Technical Trade Questions

1. There are four 0.25 g in 1g.

2. (a) $\dfrac{3}{4} = 0.75$ (c) $\dfrac{3}{8} = 0.375$

   (b) $\dfrac{5}{16} = 0.3125$ (d) $\dfrac{5}{32} = 0.15625$

3. 3.25

4. $18\dfrac{1}{4} = 18.25$ feet

   $6\dfrac{1}{3} = 6.33$ feet

   $24\dfrac{1}{8} = 24.125$ feet

5. (a) $\dfrac{3}{64}$ in.

   (b) $\dfrac{55}{64}$ in.

6. $2\dfrac{9}{16} = 2.5625$ in.
   $\phantom{2\dfrac{9}{16}} + 0.05$ in.
   $\phantom{2\dfrac{9}{16} =}\overline{2.6125 \text{ in.}}$

STEP 2-2

1. Addition, subtraction
2. (a) Subtraction   (c) Division
   (b) Addition   (d) Multiplication

3. (a) 26.443   (c)   894.357
        197.1 $\times$ 8.6
        2.1103 $\overline{5366142}$
   $+\ \ \ 0.004$ 7154856
   $\overline{225.6573}$ $\overline{7691.4702}$

   (b) 19637.224   (d)   89.31333
       $-\ 866.43$ $0.015\overline{)1.3397}$
       $\overline{18770.794}$ or $15\overline{)1339.7}$

4. $\dfrac{176 \div 8}{8 \div 8} = \dfrac{22}{1}$ or 22 : 1 or 22.0 to 1 (ratio of twenty-two to one)

5. (a) 86.44   (b) 12,263,415.01   (c) 0.18

6. 86.43760   (a) 7
              (b) 6
   12,263,415.00510   (a) 13
                     (b) 12
   0.176600   (a) 6
              (b) 4

7. (a) 0.075   (b) 220   (c) 0.235   (d) 19.2
8. (a) 35   (b) 4004

STEP 2-2 Technical Trade Questions

1. (a) AWG 11
   (b) AWG 11
   (c) $0.\overset{9}{\cancel{1}}{}^{1}02$
       -0.081
       $\overline{0.921}$
       $0.\overset{3}{\cancel{4}}{}^{1}38$

2. (a) -0.266
       $\overline{0.172}$ in.

   (b) $\overset{4}{0.406}$
       $\times7$
       $\overline{2.842}$ in.

   (c) 7-0

3. $330\ \Omega + 10\%$
   $= 330\ \Omega + 33\ \Omega = 363\ \Omega$

   $330\ \Omega - 10\%$
   $= 330\ \Omega - 33\ \Omega = 297\ \Omega$

4. 3% of 1700 = 51 bricks
   4% of 1700 = 68 bricks

5. Rounded $\dfrac{1}{1000}$
   $\overline{1.654}$
   1.665
   1.655
   1.656
   Average = 1.6575
   Rounded = 1.658

6. 6.25
   8.5
   4.5
   6.75
   $\underline{2}$
   28 hours   Average = 5.6 hours
   (0.6 × 60 minutes = 36 minutes)
   Average = 5 : 36

STEP 3-1
1. (a) $+3 - (-4) = +7$
   (b) $12 \times (+4) = +48$
   (c) $-5 \div (-7) = 0.7142$
   (d) $-0.63 \times (+6.4) = -4.032$
2. (a) $+6 + (-7)$
   (b) $-0.75 \times (18)$ or -0.75×18
3. (a) [7] [.] [5] [+/−] [+] [4] [.] [6] [+/−] [=]
   (b) [5] [×] [2] [+/−] [=]
   (c) [3] [1] [6] [.] [6] [2] [9] [÷] [1] [.] [4] [4] [+/−] [=]

STEP 3-1 Technical Trade Questions
1. (a) A = +5 V
      B = −7 V
      C = +12 V
   (b) A and B = 12 V
       B and C = 19 V
       A and C = 7 V
2. (a)
	WEEK 1	WEEK 2
M	+10	+405
T	−65	+885
W	+130	−51
T	−20	−65
F	−35	−5

   (b) Week 1 = +20, Week 2 = +1169
   (c) Week 1: Profit $20
       Week 2: Profit $1169
3. 36°C = +1.2 mm
   35°C = +0.8 mm
   34°C = +0.4 mm
   33°C = 0
   32°C = −0.4 mm
   31°C = −0.8 mm
   30°C = −1.2 mm

STEP 3-2
1. $+15 + (+4) = +19$
2. $-3 + (+45) = +42$
3. $+114 + (-111) = +3$
4. $-357 + (-74) = -431$
5. $17 + (15) = +32$
6. $-8 + (-177) = -185$
7. $+4600 + (-3400) = +1200$
8. $-6.25 + (+0.34) = -5.91$

STEP 3-2 Technical Trade Questions
1. (a) Average current relative to zero reference = +2.1 A
   (b) Average load current = 6.6 A + 2.1 A = 8.7 A
2. (a) Average temperature relative to zero reference = −2°F
   (b) Average temperature = 78°F − 2 = 76°F

STEP 3-3
1. $+18 - (+7) = +11$
2. $-3.4 - (-5.7) = +2.3$
3. $19{,}665 - (-5{,}031) = +24{,}696$
4. $-8 - (+5) = -13$
5. $467 - 223 = +244$
6. $-331 - (-2.6) = -328.4$
7. $8 - (+25) = -17$
8. $-0.64 - (-0.04) = -0.6$

STEP 3-3 Technical Trade Questions
1. 5 V ± 0.75 V = 4.25 V to 5.75 V
   −12 V ± 2.2 V = 9.8 V to 14.2 V

UNITS	A	B	C	D
+5 V	✓	✓	✗	✓
−12 V	✓	✓	✓	✓

✓ = Within Tolerance
✗ = Outside Tolerance

2. (a) 4.2 cm ± 0.3 cm = 3.9 cm to 4.5 cm
       2.6 cm ± 0.3 cm = 2.3 cm to 2.9 cm
   (b)
PART	1	2	3	4	5
A	✗	✓	✓	✗	✓
B	✓	✗	✓	✓	✓

   ✓ = Within Tolerance
   ✗ = Outside Tolerance

STEP 3-4
1. $4 \times (+3) = +12$
2. $+17 \times (-2) = -34$
3. $-8 \times (+16) = -128$
4. $-8 \times (-5) = +40$
5. $+12.6 \times (+15) = +189$
6. $-3.3 \times (+1.4) = -4.62$
7. $+0.3 \times (-4) = -1.2$
8. $-4.6 \times (-3.3) = +15.18$

STEP 3-4 Technical Trade Questions
1. $\frac{1}{3} \times (-7.6 \text{ V}) = -2.53 \text{ V}$

   The positive peak = +2.53 V

2. $°C = \frac{5}{9} \times (°F - 32)$

   (a) $\frac{5}{9} \times (16 - 32) = \frac{5}{9} \times (-16) = -8\frac{8}{9} = -8.89°C$

   (b) $\frac{5}{9} \times (76 - 32) = 24\frac{4}{9} = +24.44°C$

   (c) $\frac{5}{9} \times (4 - 32) = -15\frac{5}{9} = -15.55°C$

   (d) $\frac{5}{9} \times (98 - 32) = 36\frac{2}{3} = 36.66°C$

STEP 3-5
1. $+16.7 \div (+2.3) = 16.7 \div 2.3 = 7.26$
2. $+18 \div (+6) = 3$
3. $-6 \div (+2) = -3$
4. $+18 \div (-4) = -4.5$
5. $+2 \div (-8) = -0.25$
6. $-8 \div (+5) = -1.6$
7. $-15 \div (-5) = 3$
8. $0.664 \div (-0.2) = -3.32$

STEP 3-5 Technical Trade Questions
1. A = −180 V ÷ 3 = −60 volts
   B = −180 V ÷ 8 = −22.5 volts
2. (a) $°F = \frac{9 \times °C}{5} + 32 = \frac{9 \times -26}{5} + 32 = -14.8°F$
   (b) $°F = \frac{9 \times °C}{5} + 32 = \frac{9 \times -8}{5} + 32 = +17.6°F$

STEP 3-6
1. $+6 + (+3) + (-7) + (-5) = -3$
2. $+9 - (+2) - (-13) - (-4) = +24$
3. $-6 \times (-4) \times (-5) = -120$
4. $+8 \div (+2) \div (-5) = -0.8$
5. $-4 \div (-2) \times (+8) = +16$
6. $-9 + (+5) - (-7) = +3$

STEP 3-6 Technical Trade Questions
1. $L = x \times y \times 1.05 \times (x + 3)$
   $= 8 \times 36 \times 1.05 \times (8 + 30)$
   PEMDAS: 1. Parentheses
   　　　　　2. Multiplication
   $L = 8 \times 36 \times 1.05 \times (38) = 11{,}491$ feet
2. $C = \dfrac{3.142 \times 4.25 \text{ in.} \times 220 \text{ rpm}}{12}$
   PEMDAS: 1. Multiplication
   　　　　　2. Division
   $C = \dfrac{2937.77}{12} = 244.8$ feet/minute
3. $L = ((2 \times D) + 1.57) \times (d_1 + d_2) + \dfrac{(d_1 + d_2)}{4 \times D}$
   $= ((2 \times 12 \text{ in.}) + 1.57) \times (18 \text{ in.} + 2 \text{ in.}) + \dfrac{(18 \text{ in.} + 2 \text{ in.})}{4 \times 12}$
   PEMDAS: 1. Parentheses
   　　　　　2. Multiplication
   　　　　　3. Division
   　　　　　4. Addition
   $L = (24 + 1.57) \times 20 + \dfrac{20}{4 \times 12}$ (P)
   $L = 25.57 \times 20 + \dfrac{20}{4 \times 12}$ (P)
   $L = 45.57 + \dfrac{20}{48}$ (M)
   $L = 45.57 + 0.42$ (D)
   $L = 46$ inches (A)

STEP 4-1
1. (a) $16^4 = 16 \times 16 \times 16 \times 16 = 65{,}536$
   (b) $32^3 = 32 \times 32 \times 32 = 32{,}768$
   (c) $112^2 = 112 \times 112 = 12{,}544$
   (d) $15^6 = 15 \times 15 \times 15 \times 15 \times 15 \times 15 = 11{,}390{,}625$
   (e) $2^3 = 2 \times 2 \times 2 = 8$
   (f) $3^{12} = 3 \times 3 \times 3 \times 3 \times 3 \times 3 \times 3 \times 3 \times 3 \times 3 \times 3 \times 3 = 531{,}441$
2. (a) $\sqrt{144} = 12$ (c) $\sqrt{20} = 4.47$
   (b) $\sqrt[3]{3375} = 15$ (d) $\sqrt[3]{9} = 2.08$
3. (a) $(9^2 + 14^2)^2 - \sqrt[3]{3 \times 7}$
   $= (81 + 196)^2 - \sqrt[3]{21}$
   $= 277^2 - 2.76$
   $= 76{,}729 - 2.76 = 76{,}726.24$
   (b) $\sqrt{3^2 \div 2^2} + \dfrac{151 - 9^2}{3.5^2}$
   $= \sqrt{9 \div 4} + \dfrac{151 - 81}{12.25}$
   $= \sqrt{2.25} + \dfrac{70}{12.25}$
   $= 1.5 + 5.71 = 7.21$

STEP 4-1 Technical Trade Questions
1. $\text{BMI} = \dfrac{703 \times 180}{5'6''^2} = \dfrac{703 \times 180}{66^2} = \dfrac{126{,}540}{4356} = 29$
   Patient is overweight, bordering on obese.
2. $\text{Speed}_{(max)} = \sqrt{30 \times 0.38 \times 129} = 38$ mph
3. $Z = \sqrt{33 \, \Omega^2 + 257.8 \, \Omega^2}$
   $= \sqrt{1089 + 66460.8}$
   $= \sqrt{67549.8} = 260 \, \Omega$
4. $V = 12.14\sqrt{65} = 12.14 \times 8.1 = 98$ feet per second

STEP 4-2
1. (a) 10^2 (d) 10^6
   (b) 10^0 (e) 10^{-3}
   (c) 10^1 (f) 10^{-6}
2. (a) $6.3 \times 10^3 = 6{\underset{\frown}{.}}300. = 6300.0$ or 6300
   (b) $114{,}000 \times 10^{-3} = 114{\overset{\frown}{.}}000. = 114.0$ or 114
   (c) $7.114{,}632 \times 10^{-6} = 7{\underset{\frown}{.}}114632. = 7.114632$
   (d) $6624 \times 10^6 = 6624{\underset{\frown}{.}}000000. = 6{,}624{,}000{,}000.0$
3. (a) $\sqrt{3 \times 10^6} = \sqrt{3{,}000{,}000} = 1732.05$
   (b) $(2.6 \times 10^{-6}) - (9.7 \times 10^{-9}) = 0.0000025$ or 2.5×10^{-6}
   (c) $\dfrac{(4.7 \times 10^3)^2}{3.6 \times 10^6} = (4.7 \times 10^3)^2 \div (3.6 \times 10^6) = 6.14$
4. (a) $47{,}000 = 47{\overset{\frown}{.}}000. = 47 \times 10^3$
   (b) $0.00000025 = 0{\underset{\frown}{.}}00000025\overset{\frown}{0.} = 250 \times 10^{-9}$
   (c) $250{,}000{,}000 = 250{\overset{\frown}{.}}000000. = 250 \times 10^6$
   (d) $0.0042 = 0{\underset{\frown}{.}}004.2 = 4.2 \times 10^{-3}$

STEP 4-2 Technical Trade Questions
1. (a) $98{,}000$ volts $= 98$ kilovolts
   (b) 0.18 amps $= 180$ milliamps
   (c) $166{,}000{,}000$ volts $= 166$ megavolts
   (d) 3300 ohms $= 3.3$ kiloohms
2. $\dfrac{6400}{0.2 \times 10^3} = 32$ gallons
3. This should not be confused with a power of 10 exponent (d_L/d_S to the fifth power)
   $N = \sqrt{\left(\dfrac{d_L}{d_S}\right)^5} = \sqrt{\left(\dfrac{1.5}{0.75}\right)^5} = \sqrt{32} = 5.7$ pipes (6 pipes)
4. $C = \dfrac{(8.85 \times 10^{-12}) \times x \times A}{d}$
   $= \dfrac{(8.85 \times 10^{-12}) \times 6 \times 0.3}{0.0003}$
   $= 5.31 \times 10^{-8}$ or $0.0531 \, \mu\text{F}$ (microfarads)

STEP 4-3
1. (a) 10^3 (d) 10^6
   (b) 10^{-2} (e) 10^{-6}
   (c) 10^{-3}
2. (a) cm $\times 0.4 =$ inches, 15 cm $\times 0.4 = 6$ inches
   (b) kg $\times 2.2 =$ pounds, 23 kg $\times 2.2 = 50.6$ pounds
   (c) liters $\times 0.26 =$ gallons, 37 L $\times 0.26 = 9.62$ gallons
   (d) $\left(\dfrac{9}{5} \times °\text{C}\right) + 32 = \left(\dfrac{9}{5} \times 23\right) + 32 = 73.4°\text{F}$
3. (a) Miles/hour $\times 1.6 =$ kilometers/hour,
       55 mph $\times 1.6 = 88$ km/hour
   (b) gallons $\times 3.8 =$ liters, 16 gallons $\times 3.8 = 60.8$ liters
   (c) square yards $\times 0.8 =$ square meters,
       3 yd$^2 \times 0.8 = 2.4$ m^2
   (d) $\dfrac{5}{9} \times (°\text{F} - 32) = \dfrac{5}{9} \times (92 - 32) = 33.33°\text{C}$
4. (a) meter (m)   (f) ampere (A)
   (b) gram (g)   (g) volt (V)
   (c) degree Celsius (°C) (h) liter (L)
   (d) second (s)   (i) joule (J)
   (e) watt (W)   (j) ohm (Ω)

5. (a) 25,000 volts = 25.000 × 10 volts
   = 25 kilovolts

   (b) 0.014 watts = 0.014. = 14 × 10⁻³ watts
   = 14 milliwatts

   (c) 0.000016 microfarad = 0.000016 × 10⁻⁶ farad
   = 0.000 016 × 10⁻⁹ farad
   = 0.016 × 10⁻⁹ farad
   = 0.016 nanofarad

STEP 4-3 Technical Trade Questions

1. Length = 8 feet (should be in feet, so okay)
   Width = $3\frac{1}{2}$ inches (should be in feet, so must be converted)
   Height = $1\frac{1}{2}$ inches (should be in inches, so okay)

   $3\frac{1}{2}$ inches of 12 inches

   $\dfrac{3\frac{1}{2}}{12} = 3\frac{1}{2} \div 12$

   $= \dfrac{7}{2} \div \dfrac{12}{1}$

   $= \dfrac{7}{2} \times \dfrac{1}{12} = \dfrac{7}{24}$

   $N = l \times w \times h$
   $= 8 \times \frac{7}{24} \times 1\frac{1}{2} = 3\frac{1}{2}$ board feet
   $3\frac{1}{2}$ bf

2. 11×1 mm = 11.00
   0×0.5 mm = 0.00
   16×0.01 mm = 0.16
   ─────
   11.16 mm

3. A (Ammeter) is monitoring current.
   B (Voltmeter) is measuring voltage.
   $P = V \times I$
   = 13.6 V × 970 mA
   = 13.2 watts

4. There are 2.54 cm in 1 inch, and so:
   There are 25.4 mm in 1 inch.
   $2\frac{1}{4} \times 25.4 = 57.15$ mm
   $2\frac{3}{4} \times 25.4 = 69.85$ mm

5. 600 cu. in. × 16.387 = 9,832.2 cm³
   since 1 cm³ = 1 mL
   9,832.2 cm³ = 9,832.2 mL
   9,832.2 mL ÷ 1000 = 9.8 Liters
   (rounded) = 10 Liter engine

6. 8 feet =
   × 4 feet =
   × 4 feet =
   There are 30 cm in 1 foot.
   8 × 30 = 240 cm
   4 × 30 = 120 cm
   Metric cord = 240 cm by 120 cm × 120 cm
   (100 cm = 1 meter)
   2.4 m by 1.2 m × 1.2 m

STEP 5-1

1. Yes
2. No
3. $\dfrac{144}{12} \times \square = \dfrac{36}{6} \times 2 \times \square = 60$

   $12 \times \square = 6 \times 2 \times \square = 60$

   $12 \times 5 = 12 \times 5 = 60$

   $\square = 5$

4. $\dfrac{(8-4) + 26}{5} = \dfrac{81 - 75}{2}$

   $\dfrac{4 + 26}{5} = \dfrac{6}{2}$

   $\dfrac{30}{5} = \dfrac{6}{2}$

   $6 = 3$ (equation is not equal)

5. Yes

STEP 5-1 Technical Trade Questions

1. (a) $\dfrac{L_1}{L_2} = \dfrac{R_1}{R_2}$

   $\dfrac{18}{45} = \dfrac{3}{7.5}$

   $0.4 = 0.4$
   Yes

   (b) L_1 is doubled from 18′ to 36′.
   R_1 is doubled from 3 Ω to 6 Ω.
   $\dfrac{36}{45} = \dfrac{6}{7.5}$
   $0.8 = 0.8$
   Yes

2. (a) $\dfrac{S_x}{S_y} = \dfrac{d_y}{d_x}$

   $\dfrac{120}{480} = \dfrac{4}{16}$

   $0.25 = 0.25$
   Yes

   (b) d_x is halved from 16″ to 8″.
   S_x is doubled from 120 rpm to 240 rpm.
   $\dfrac{240}{480} = \dfrac{4}{8}$
   $0.5 = 0.5$
   Yes

STEP 5-2

1. (a) $x + x = 2x$
   (b) $x \times x = x^2$
   (c) $7a + 4a = 11a$
   (d) $2x - x = 1x$ or x
   (e) $\dfrac{x}{x} = 1$
   (f) $x - x = 0$

2. (a) $x + 14 = 30$
   $x + 14 - 14 = 30 - 14$ (−14 from both sides)
   $x = 30 - 14$
   $x = 16$

   (b) $8 \times x = \dfrac{80 - 40}{10} \times 12$
   $8 \times x = 4 \times 12$
   $8 \times x = 48$
   $\dfrac{8 \times x}{8} = \dfrac{48}{8}$ (÷8)
   $x = \dfrac{48}{8}$
   $x = 6$

   (c) $y - 4 = 8$
   $y - 4 + 4 = 8 + 4$ (+4)
   $y = 8 + 4$
   $y = 12$

   (d) $(x \times 3) - 2 = \dfrac{26}{2}$
   $(x \times 3) - 2 + 2 = \dfrac{26}{2} + 2$ (+2)

$$x \times 3 = \frac{26}{2} + 2$$
$$x \times 3 = 15$$
$$\frac{x \times 3}{3} = \frac{15}{3} \quad (\div 3)$$
$$x = \frac{15}{3}$$
$$x = 5$$

(e)
$$x^2 + 5 = 14$$
$$x^2 + 5 - 5 = 14 - 5 \quad (-5)$$
$$x^2 = 14 - 5$$
$$\sqrt{x^2} = \sqrt{14 - 5} \quad (\sqrt{\ })$$
$$x = \sqrt{14 - 5}$$
$$x = \sqrt{9}$$
$$x = 3$$

(f)
$$2(3 + 4x) = 2(x + 13)$$
$$6 + 8x = 2x + 26 \quad \text{(remove parentheses)}$$
$$6 + 8x - 2x = 2x + 26 - 2x \quad (-2x)$$
$$6 + (8x - 2x) = 26$$
$$6 + 6x = 26$$
$$6 + 6x - 6 = 26 - 6 \quad (-6)$$
$$6x = 26 - 6$$
$$6x = 20$$
$$\frac{6 \times x}{6} = \frac{20}{6} \quad (\div 6)$$
$$x = \frac{20}{6}$$
$$x = 3.3333$$

3. (a) $x + y = z, y = ?$
$$x + y - x = z - x \quad (-x)$$
$$y = z - x$$

(b) $Q = C \times V, C = ?$
$$\frac{Q}{V} = \frac{C \times V}{V} \quad (\div V)$$
$$\frac{Q}{V} = C$$
$$C = \frac{Q}{V}$$

(c) $X_L = 2 \times \pi \times f \times L, L = ?$
$$\frac{X_L}{2 \times \pi \times f} = \frac{2 \times \pi \times f \times L}{2 \times \pi \times f} \quad (\div 2 \times \pi \times f)$$
$$\frac{X_L}{2 \times \pi \times f} = L$$
$$L = \frac{X_L}{2 \times \pi \times f}$$

(d) $V = I \times R, R = ?$
$$\frac{V}{I} = \frac{I \times R}{I} \quad (\div I)$$
$$\frac{V}{I} = R$$
$$R = \frac{V}{I}$$

4. (a)
$$I^2 = 9$$
$$\sqrt{I^2} = \sqrt{9} \quad (\sqrt{\ })$$
$$I = \sqrt{9}$$
$$I = 3$$

(b)
$$\sqrt{Z} = 8$$
$$\sqrt{Z}^2 = 8^2$$
$$Z = 8^2$$
$$Z = 64$$

STEP 5-2   Technical Trade Questions

1. (a) $H = \dfrac{T \times V}{375} = \dfrac{120{,}000 \times 450}{375} = 14{,}400$

   (b) $V = \dfrac{375 \times H}{T}$

   (c) $V = \dfrac{375 \times 14{,}400}{12{,}000} = 450$

2. (a) $C_A = 8.34 \times W \times C_c$
   $= 8.34 \times 4.5 \times 24.3 = 912$ pounds

   (b) $C_C = \dfrac{C_A}{8.34 \times W}$

   (c) $C_C = \dfrac{912}{8.34 \times 4.5} = 24.3$

3. (a) $C_D = \dfrac{C_A \times A_D}{C_A + 12} = \dfrac{8 \times 125}{8 + 12} = 50$ mg

   (b) $A_D = \dfrac{C_D(C_A + 12)}{C_A}$

   (c) $A_D = \dfrac{50(8 + 12)}{8} = 125$ mg

4. (a) $L_T = L_1 + L_2 + \left(\dfrac{1}{2} \times T\right)$
   $= 10\,\text{cm} + 5\,\text{cm} + \left(\dfrac{1}{2} \times 0.5\,\text{cm}\right)$
   $= 15.25\,\text{cm}$

   (b) 0.25 cm

   (c) $L_1 = L_T - L_1 - \left(\dfrac{1}{2} \times T\right)$
   $L_2 = L_T - L_1 - \left(\dfrac{1}{2} \times T\right)$

STEP 5-3

1. $x = y \times z$ and $a = x \times y, y = 14, z = 5, a = ?$
$$a = x \times y$$
$$a = y \times z \times y \quad \text{(substitute } y \times z \text{ for } x)$$
$$a = y^2 \times z \quad (y \times y = y^2)$$
$$a = 14^2 \times 5$$
$$a = 980$$

2. (a) For $I = \sqrt{\dfrac{P}{R}}$,
$$P = V \times I \text{ to } P = I^2 \times R$$
$$\frac{P}{R} = \frac{I^2 \times R}{R} \quad (\div R)$$
$$\frac{P}{R} = I^2$$
$$\sqrt{\frac{P}{R}} = \sqrt{I^2} \quad (\sqrt{\ })$$
$$\sqrt{\frac{P}{R}} = I$$
$$I = \sqrt{\frac{P}{R}}$$

(b) For $I = \dfrac{P}{V}$,
$$P = V \times I$$
$$\frac{P}{V} = \frac{V \times I}{V} \quad (\div V)$$
$$\frac{P}{V} = I$$
$$I = \frac{P}{V}$$

(c) For $V = \sqrt{P \times R}$

$$P = V \times I \text{ to } P = \frac{V^2}{R}$$

$$P \times R = \frac{V^2 \times \cancel{R}}{\cancel{R}} \quad (\times R)$$

$$P \times R = V^2$$

$$\sqrt{P \times R} = \sqrt{V^2} \quad (\sqrt{})$$

$$\sqrt{P \times R} = V$$

$$V = \sqrt{P \times R}$$

(d) For $R = \frac{V^2}{P}$,

$$P = V \times I \text{ to } P = \frac{V^2}{R}$$

$$P \times R = \frac{V^2}{\cancel{R}} \times \cancel{R} \quad (\times R)$$

$$P \times R = V^2$$

$$\frac{\cancel{R} \times R}{\cancel{R}} = \frac{V^2}{P}$$

$$R = \frac{V^2}{P}$$

3. What percentage of 12 gives 2.5?
$$x \times 12 = 2.5$$
$$\frac{x \times \cancel{12}}{\cancel{12}} = \frac{2.5}{12} \quad \div 12$$
$$x = \frac{2.5}{12}$$
$$x = 0.2083 \quad (0.2083 \times 12 = 2.5)$$
$$x\% = 20.83\% \quad (20.83\% \times 12 = 2.5)$$

4. 60 is 22% of what number?
$$60 = 22\% \times x$$
$$60 = 0.22 \times x$$
$$\frac{60}{0.22} = \frac{\cancel{0.22} \times x}{\cancel{0.22}} \quad (\div 0.22)$$
$$\frac{60}{0.22} = x$$
$$x = 272.73 \quad (60 = 0.22 \times 272.73$$
$$\text{or } 60 = 22\% \times 272.73)$$

STEP 5-3 Technical Trade Questions

1. (a) $P \approx \frac{T^2 \times d \times N}{3.78}$ Substitute d with $2 \times r$ (diameter = twice the radius)

$$P \approx \frac{T^2 \times (2 \times r) \times N}{3.78}$$

(b) $N \approx \frac{3.78 \times P}{d \times T^2}$

(c) $P \approx \frac{T^2 \times d \times N}{3.78} = \frac{3^2 \times 0.5 \times 4}{3.78} = 4.76 \text{ hp}$

2. (a) $W_T = 0.5(d_O - d_I)$ (substitute d_I with $\frac{2}{3} \times d_O$)

$$W_T = 0.5\left(d_O - \left(\frac{2}{3} \times d_O\right)\right)$$

(b) $W_T = 0.5\left(2.25 - \left(\frac{2}{3} \times 2.25\right)\right)$
$$= 0.5(2.25 - 1.5)$$
$$= 0.5 \times 0.75$$
$$= 0.375 \text{ in.}$$

(c) $W_T = \frac{d_O}{6} = \frac{2\frac{1}{4}}{6} = \frac{3}{8}'' \leftarrow 2\frac{1}{4} \div 6 = \frac{9}{4} \div \frac{6}{1}$
$$= \frac{9}{4} \times \frac{1}{6} = \frac{9}{24} = \frac{3}{8}$$
$$= 0.375''$$

STEP 5-4

1. $y = -2x + 3$
2. $(x - 5)(x - 3)$
3. $x = 0$ or $x = -3$
4. $16x^8y^4$

STEP 5-4 Technical Trade Questions

1. (a) Period $= \frac{1}{\text{freq.}} = \frac{1}{1 \times 10^3} = 1 \times 10^{-3}$ seconds,

or 0.001 second $\left(\frac{1}{1000}\text{th of a second}\right)$

Period $= \frac{1}{\text{freq.}} = \frac{1}{24 \times 10^3} = 4.2 \times 10^{-5}$ seconds,

or 0.000042 second

(b) $f = \frac{1}{P} = \frac{1}{1 \times 10^3} = 1 \times 10^3$ hertz, or 1000 Hz. (1000 cycles every second)

$f = \frac{1}{P} = \frac{1}{560 \times 10^{-6}} = 1785.7$ Hertz, or 1.8 kilohertz

2. (a) $P = 62.4 \times D = 62.4 \times 6 = 374.4$ pounds per square foot

(b) $D = \frac{P}{62.4} = \frac{1200}{62.4} = 19.2$ feet

3. (a) $H_{R(max)} = (0.8 \times A) + 176 = (-0.8 \times 40) + 176 = 144$
140 is below the maximum target heart rate of 144.

(b) $A = 220 - 1.25 \times H_{R(max)}$
$= 220 - 1.25 \times 160$
$= 220 - 200$
$= 20$ years old

STEP 5-5

1. Parentheses, Exponents, Multiplication, Division, Addition, Subtraction.
2. Monomial: $3x$
Binomial: $3x + 4y$
Trinomial: $3x + 4y + 5z$
Polynomial: $2a + 2b + 2c + 2d + 2e$
3. (a) $a \div 1 = a$
   (b) $a + a = 2a$
   (c) $a - a = 0$
   (d) $a \times a = a^2$
   (e) $3x + b - 2x = x + b$
4. (a) $16 + a = 5 \times a$
$a = 4$

(b) $\frac{42}{b} = 3 \times 7$
$b = 2$

STEP 6-1

1. (a) 40°   (d) 90°   (g) 165°   (j) 165°
   (b) 60°   (e) 115°   (h) 180°   (k) 65°
   (c) 80°   (f) 135°   (i) 15°   (l) 90°

2. (a) Acute   (d) Right   (g) Obtuse   (j) Obtuse
   (b) Acute   (e) Obtuse   (h) Obtuse   (k) Acute
   (c) Acute   (f) Obtuse   (i) Acute   (l) Right

3. (See Figure 6-1)

STEP 6-1 Technical Trade Questions
1. 100°
2. (a) 90°
   (b) 135°
3. 30°
4. 55°

STEP 6-2
1. (Section 6-2 Introduction)
2. (a) Figure 6-2 (e) Figure 6-6 (i) Figure 6-12
   (b) Figure 6-3 (f) Figure 6-7 (j) Figure 6-13
   (c) Figure 6-4 (g) Figure 6-10
   (d) Figure 6-5 (h) Figure 6-11
3. (a) $A = \sqrt{C^2 - B^2}$
      $= \sqrt{3^2 - 2^2}$
      $= \sqrt{9 - 4}$
      $= \sqrt{5}$
      $= 2.24$ feet
   (b) $B = \sqrt{C^2 - A^2}$
      $= \sqrt{160^2 - 80^2}$
      $= \sqrt{25{,}600 - 6400}$
      $= \sqrt{19{,}200}$
      $= 138.56$ km
   (c) $C = \sqrt{A^2 + B^2}$
      $= \sqrt{112^2 + 25^2}$
      $= \sqrt{12{,}544 + 625}$
      $= \sqrt{13{,}169}$
      $= 114.76$ mm
4. (a) $x = \sqrt{A^2 + B^2}$
      $= \sqrt{40^2 + 30^2}$
      $= \sqrt{1600 + 900}$
      $= \sqrt{2500}$
      $= 50$ volts
   (b) $x = \sqrt{A^2 + B^2}$
      $= \sqrt{75^2 + 26^2}$
      $= \sqrt{5625 + 676}$
      $= \sqrt{6301}$
      $= 79.38$ watts
   (c) $x = \sqrt{A^2 + B^2}$
      $= \sqrt{93^2 + 36^2}$
      $= \sqrt{8649 + 1296}$
      $= \sqrt{9945}$
      $= 99.72$ mm
      $y = \sqrt{A^2 + B^2}$
      $= \sqrt{48^2 + 96^2}$
      $= \sqrt{2304 + 9216}$
      $= \sqrt{11{,}520}$
      $= 107.33$ mm

STEP 6-2 Technical Trade Questions
1. Area $(A) = $ length $(l) \times$ width (w)
   $= 6.2$ cm $\times 1.3$ cm $= 8.06$ cm^2
2. $18' \times 12' = 216$ square feet $(216$ ft$^2)$
   25% of $216 = 54$ ft^2
   $= 270$ ft^2
   $270 \div 16 = 16.9$ (17 packs)

3. $A = \dfrac{a \times b}{2} = \dfrac{40 \times 30}{2} = 600$ ft^2
   $C = \sqrt{a^2 + b^2} = \sqrt{40^2 + 30^2} = \sqrt{1600 + 900} = \sqrt{2500} = 50$ feet

4. (a) Large circle: $c = 2$ cm, $r = 1$ cm,
      $A = \pi r^2 = \pi \times 1^2 = 3.142$ cm^2
      Small circle: $c = 0.5$ cm, $r = 0.25$ cm,
      $A = \pi r^2 = \pi \times 0.25^2 = 0.19$ cm^2
      Total area: $3.142 - 0.19 = 2.95$ cm^2
   (b) Half circle: $r = 0.5''$, $A = \pi r^2 = \pi \times 0.5^2 = 0.79$,
      $0.79 \div 2 = 0.4$ in^2
      Rectangle: $A = l \times w = 1.5 \times 0.5 = 0.75$ in^2
      Total area: $0.75 + 0.4 = 1.15$ in^2
   (c) Circle area: $A = \pi r^2 = \pi \times 2.5^2 = 19.6$ cm^2
      Hexagon area: $A = \dfrac{b \times h}{2} \times 6 = \dfrac{4 \text{ cm} \times 5 \text{ cm}}{2} \times 6 = 60$ cm^2
      Total area: $60 - 19.6 = 40.4$ cm^2
   (d) Circle area: $A = \pi r^2 = \pi \times 2.2^2 = 15.2$ cm$^2 \div 2 = 7.6$ cm^2
      Trapezoid area:
      (Upper base) $= 1.2$ cm $+ 4.4$ cm $+ 1.4$ cm $= 7$ cm
      (Lower base) $= 9.8$ cm
      $A = \dfrac{h}{2} \times (B + b)$
      $= \dfrac{5.4 \text{ cm}}{2} \times (9.8 \text{ cm} + 7 \text{ cm})$
      $= 2.7$ cm $\times 16.8$ cm $= 45.36$ cm^2

STEP 6-3
1. (a) $O = 35$ mm $\theta = 36°$
      $H = ?$

   SOH   or $\sin \theta = \dfrac{O}{H}$
   $\sin 36° = \dfrac{35 \text{ mm}}{H}$
   $0.59 = \dfrac{35}{H}$
   $0.59 \times H = \dfrac{35}{H} \times H$ $(\times H)$
   $\dfrac{0.59 \times H}{0.59} = \dfrac{35}{0.59}$ $(\div 0.59)$
   $H = \dfrac{35}{0.59} = 59.32$ mm

   (b) $H = 160$ km $\theta = 38°$
      $A = ?$

   CAH   or $\cos \theta = \dfrac{A}{H}$
   $\cos 38° = \dfrac{A}{160 \text{ km}}$
   $0.79 = \dfrac{A}{160}$
   $0.79 \times 160 = \dfrac{A}{160} \times 160$ $(\times 160)$
   $0.79 \times 160 = A$
   $A = 126.4$ km

(c) $O = 163$ cm $\quad\quad \theta = 72°$
$A = ?$

$\quad$ TOA $\quad$ or $\quad\quad \tan\theta = \dfrac{O}{A}$

$\quad\quad\quad\quad\quad\quad \tan 72° = \dfrac{163 \text{ cm}}{A}$

$\quad\quad\quad\quad\quad\quad 3.08 = \dfrac{163}{A}$

$\quad\quad\quad\quad\quad\quad 3.08 \times A = \dfrac{163 \times \cancel{A}}{\cancel{A}} \quad\quad (\times A)$

$\quad\quad\quad\quad\quad\quad \dfrac{3.08 \times A}{3.08} = \dfrac{163}{3.08} \quad\quad (\div 3.08)$

$\quad\quad\quad\quad\quad\quad A = \dfrac{163}{3.08} = 52.92$ cm

2. (a) $H = 120$ miles $\quad\quad \theta = ?$
$\quad\quad O = 38$ miles

$\quad\quad$ SOH $\quad$ or $\quad\quad \sin\theta = \dfrac{O}{H}$

$\quad\quad\quad\quad\quad\quad \sin\theta = \dfrac{38 \text{ miles}}{120 \text{ miles}}$

$\quad\quad\quad\quad\quad\quad \sin\theta = 0.32$

$\quad\quad\quad\quad\quad\quad \cancel{\text{invsin}} \times \cancel{\sin}\theta = \text{invsin } 0.32 \quad (\times \text{invsin})$

$\quad\quad\quad\quad\quad\quad \theta = \text{invsin } 0.32$

$\quad\quad\quad\quad\quad\quad \theta = 18.66°$

(b) $H = 25$ feet $\quad\quad \theta = ?$
$\quad\quad A = 17$ feet

$\quad\quad$ CAH $\quad$ or $\quad\quad \cos\theta = \dfrac{A}{H}$

$\quad\quad\quad\quad\quad\quad \cos\theta = \dfrac{17 \text{ feet}}{25 \text{ feet}}$

$\quad\quad\quad\quad\quad\quad \cos\theta = 0.68$

$\quad\quad\quad\quad\quad\quad \cancel{\text{invcos}} \times \cancel{\cos}\theta = \text{invcos } 0.68$

$\quad\quad\quad\quad\quad\quad \theta = \text{invcos } 0.68$

$\quad\quad\quad\quad\quad\quad \theta = 47.16°$

(c) $A = 69$ cm $\quad\quad \theta = ?$
$\quad\quad O = 51$ cm

$\quad\quad$ TOA $\quad$ or $\quad\quad \tan\theta = \dfrac{O}{A}$

$\quad\quad\quad\quad\quad\quad \tan\theta = \dfrac{51 \text{ cm}}{69 \text{ cm}}$

$\quad\quad\quad\quad\quad\quad \tan\theta = 0.74$

$\quad\quad\quad\quad\quad\quad \cancel{\text{invtan}} \times \cancel{\tan}\theta = \text{invtan } 0.74$

$\quad\quad\quad\quad\quad\quad \theta = \text{invtan } 0.74$

$\quad\quad\quad\quad\quad\quad \theta = 36.5°$

STEP 6-3 Technical Trade Questions

1. (a) $P_A = \sqrt{P_R^2 + P_x^2} = \sqrt{12^2 + 14^2} = \sqrt{144 + 196} = \sqrt{340} = 18.44$

   (b) Opposite $= P_x = 12$ (All sides are known, so you can use either sine, cosine, or tangent)
   Hypotenuse $= P_A = 18.44$
   Adjacent $= P_R = 14$
   Sine $= \theta = \dfrac{O}{H}$
   $\theta = \text{invsine } \dfrac{O}{H} = \text{invsine } \dfrac{12}{18.44} = 40.6°$

   (c) $\theta = \text{invcosine } \dfrac{A}{H} = \text{invcosine } \dfrac{14}{18.44} = 40.6°$
   $\theta = \text{invtangent } \dfrac{O}{A} = \text{invtangent } \dfrac{12}{14} = 40.6°$

2. Hypotenuse $= P$
   Adjacent $= 22$ feet
   Angle $\theta = 18°$
   Cosine $18° = \dfrac{A}{H} = \dfrac{22'}{H}$
   $0.95 = \dfrac{22}{A}, H = \dfrac{22}{0.95} = 23.2$ feet

3. (a) $H = \sqrt{6.2^2 + 3.4^2} = \sqrt{38.44 + 11.56} = \sqrt{50} = 7.07$ cm

   (b) $\theta = \text{invsine } \dfrac{O}{H} = \text{invsine } \dfrac{3.4 \text{ cm}}{7.07 \text{ cm}} = 28.7°$

STEP 6-4

1. (Section 6-4 Introduction)
2. (Section 6-4-1)
3. (a) Figure 6-22 $\quad\quad$ (d) Figure 6-25
   (b) Figure 6-23 $\quad\quad$ (e) Figure 6-26
   (c) Figure 6-24
4. A frustum is the base section of a solid pyramid or cone.

STEP 6-4 Technical Trade Questions

1. Volume $= (x \times y) \times h$
   $= (2.25 \times 2.25) \times 4.5$
   $= 22.8$ cubic inches
   Weight $=$ volume $\times$ density
   $= 22.8 \times 0.28 = 6.38$ pounds

2. Lateral surface area $= P \times h$
   $= 72 \times 20$
   $= 1440$ square feet
   7 windows $\times (4 \times 4) = 112$ square feet
   2 doors $\times (7 \times 3) = 42$ square feet
   $1440 - 112 - 42 = 1286$ square feet
   1286 square feet $\div 300$ square feet/gallon $= 4.3$ gallons

3. Cylinder volume
   $V = 0.7854 \times d^2 \times h$
   $= 0.7854 \times 35^2 \times 1.5$
   $= 0.7854 \times 1225 \times 1.5$
   $= 1443$ cubic feet

4. Lateral surface area $= \left(\dfrac{1}{2}\right) p \times s$
   $= \left(\dfrac{1}{2}\right) 160 \times 24$
   $= \left(\dfrac{1}{2}\right) \times 3840$
   $= 1920$ square meters
   perimeter $= 4 \times 40$ meters
   $= 160$ meters
   Total cost $= 1920 \times \$5.75 = \$11,040$

5. Two $\dfrac{1}{2}$ spheres $= 1$ sphere
   Volume of sphere $= 0.524 \times d^2$
   $= 0.524 \times 2^2$
   $= 0.524 \times 4 = 2.1$ cubic inches
   Volume of cylinder $= 0.7854 \times d^2 \times h$
   $= 0.7854 \times 2^2 \times 4$
   $= 12.6$ cubic inches
   Total volume $= 2.1 + 12.6 = 14.7$ cubic inches
   Weight $=$ density $\times$ volume
   $= 0.093 \times 14.7 = 1.37$ pounds

Answers to Odd-Numbered Problems

APPENDIX B

Chapter 1
1. b
3. c
5. b
7. d
9. c
11. c

(The answers to Communication Skill Questions 13 through 23 can be found in the indicated sections that follow the questions.)

25. (a) 276 (b) 8070 (c) 20,049
27. (a) $3\frac{3}{4}$ (b) $2\frac{1}{3}$
29. (a) $\frac{5}{8} + \frac{2}{8} = \frac{5+2}{8} = \frac{7}{8}$
    (b) $\frac{47}{76} + \frac{15}{76} + \frac{1}{76} = \frac{63}{76}$
31. (a) $\frac{5}{2} = ?$, 2 into 5 = 2 with 1 remaining $\frac{5}{2} = 2\frac{1}{2}$
    (b) $\frac{17}{4} = 4\frac{1}{4}$
    (c) $\frac{25}{16} = 1\frac{9}{16}$
    (d) $\frac{37}{3} = 12\frac{1}{3}$
33. (a) $\frac{4 \div 4}{16 \div 4} = \frac{1}{4}$ (c) $\frac{74 \div 2}{128 \div 2} = \frac{37}{64}$
    (b) $\frac{16 \div 2}{18 \div 2} = \frac{8}{9}$ (d) $\frac{28}{45} = \frac{28}{45}$
35. (a) $\frac{1}{9} \times 4 = \frac{1 \times 4}{9 \times 1} = \frac{4}{9}$
    (b) $\frac{3}{6} \times \frac{4}{5} = \frac{3 \times 4}{6 \times 5} = \frac{12 \div 6}{30 \div 6} = \frac{2}{5}$
    (c) $\frac{1}{3} \times 4\frac{1}{2} = \frac{1}{3} \times \frac{9}{2} = \frac{9}{6} = 1\frac{3}{6} = 1\frac{1}{2}$
    (d) $2\frac{3}{4} \times 4\frac{4}{11} = \frac{\text{⑪}}{\text{④}} \times \frac{\text{㊽}}{\text{⑪}} = \frac{1}{1} \times \frac{12}{1} = 1 \times 12 = 12$
37. (a) $\frac{3}{4}$ (b) $\frac{1}{4}$
39. A = 5/12
    B = 3/7
41. 7/8 watt
43. 5/32, 3/16, 1/4, 3/8
45. 57 cubic yards

Chapter 2
1. a
3. b
5. a
7. c

(The answers to Communication Skill Questions 9 through 17 can be found in the indicated sections that follow the questions.)

19. (a) 2.3 (b) 0.507 (c) 9030.04
21. (a) $\frac{16}{32} \div \frac{2}{2} = \frac{1}{2} = 1 \div 2 = 0.5$
    (b) $3\frac{8}{9} = 3 + (8 \div 9) = 3 + 0.889 = 3.889$
    (c) $4\frac{9}{8} = 4 + \frac{8}{8} + \frac{1}{8} = 4 + 1 + \frac{1}{8} = 5\frac{1}{8} = 5 + (1 \div 8)$
        $= 5 + 0.125 = 5.125$
    (d) $195\frac{7}{3} = 195 + \frac{3}{3} + \frac{3}{3} + \frac{1}{3} = 197\frac{1}{3} = 197 + (1 \div 3)$
        $= 197 + 0.333 = 197.333$
23. (a) $0.777 = \frac{7}{9}$
    (b) $0.6149069 = \frac{99}{161}$
    (c) $43.125 = 43\frac{1}{8}$
25. (a) $\frac{20 \text{ ft}}{5 \text{ ft}}$ ← Reduce to lowest terms
        $\frac{20 \div 5}{5 \div 5} = \frac{4}{1}$
        The ratio of 20 ft to 5 ft is 4 to 1 (4:1).
    (b) Both quantities must be alike, so we must first convert minutes to seconds.
        $2\frac{1}{2}$ min $= 2\frac{1}{2} \times 60$ s $= 150$ s
        $\frac{150 \text{ s} \div}{30 \text{ s} \div} = \frac{5}{1}$
        The ratio of $2\frac{1}{2}$ min to 30 s is 5 to 1 (5:1).
27. 10.9
29. (a) 6 (d) 505.75, or $505\frac{3}{4}$ Ω
    (b) 17.25 or $17\frac{1}{4}$ s (e) 155 m
    (c) 116 V
31. Plugs 2 and 3
33. $6675 \times 0.25 = 1668.75$
35. $2 \times 3 \times 16 = 96$
37. $2.9 - 0.4 = 2.5$ cm, 0.4 cm
39. $0.012 - 0.0023 = 0.0097$ inch

Chapter 3
1. c
3. a
5. a
7. b
9. a

(The answers to Communication Skill Questions 11 through 20 can be found in the indicated sections that follow the questions.)

21. (a) $+8 - (-5)$ (c) $-22.3 \div (-17)$
    (b) -0.6×13 (d) $(+4) \div (-9)$
23. (a) $+6 - (+8) = -2$ (c) $-75 - (+62) = -137$
    (b) $+9 - (-6) = +15$ (d) $-39 - (-112) = +73$
25. (a) $+19 \div (+3) = 6.33$ (c) $-80 \div (+5) = -16$
    (b) $+36 \div (-3) = -12$ (d) $-44 \div (-2) = 22$
27. $+15 \div (+5) \times (-3.5) = -10.5$

215

29. $-6 \div (-4) \div (-3) \times (-15) = 7.5$
31. (a) $+3$ (c) -1
     (b) $+4$ (d) -3
33. $75 + 17 = -92$ feet
35. -4.5
37. $261°F$
39. Polytech

Chapter 4
1. c
3. a
5. d
7. a
9. b

(The answers to Communication Skill Questions 11 through 20 can be found in the indicated sections that follow the questions.)

21. (a) $9^2 = 9 \times 9 = 81$
     (b) $6^2 = 6 \times 6 = 36$
     (c) $2^2 = 2 \times 2 = 4$
     (d) $0^2 = 0 \times 0 = 0$
     (e) $1^2 = 1 \times 1 = 1$
     (f) $12^2 = 12 \times 12 = 144$
23. (a) $9^3 = 9 \times 9 \times 9 = 729$
     (b) $10^4 = 10 \times 10 \times 10 \times 10 = 10{,}000$
     (c) $4^6 = 4 \times 4 \times 4 \times 4 \times 4 \times 4 = 4096$
     (d) $2.5^3 = 2.5 \times 2.5 \times 2.5 = 15.625$
25. (a) $\dfrac{1}{100} = 10^{-2}$ (c) $\dfrac{1}{1000} = 10^{-3}$
     (b) $1{,}000{,}000{,}000 = 10^9$ (d) $1000 = 10^3$

27.

	Scientific Notation Base of 1 to 10, plus power of ten	Engineering Notation Base greater than 1, plus × 3 power of ten
(a) 475	$= 4.75 \times 10^2$	475.0
(b) 8200	$= 8.2 \times 10^3$	8.2×10^3
(c) 0.07	$= 7 \times 10^{-2}$	70×10^{-3}
(d) 0.00045	$= 4.5 \times 10^{-4}$	450×10^{-6}

29. (a) km $\times 0.6 =$ mi, 100 km $\times 0.6 = 60$ mi
     (b) m$^2 \times 1.2 =$ yd^2, 29 m$^2 \times 1.2 = 34.8$ yd^2
     (c) kg $\times 2.2 =$ lb, 67 kg $\times 2.2 = 147.4$ lb
     (d) L $\times 2.1 =$ pt, 2 L $\times 2.1 = 4.2$ pt
31. (a) $\dfrac{5}{9} \times (°F - 32) = °C$

$= \dfrac{5}{9} \times (32°F - 32) = 0.5555 \times 0 = 0°C$

$32°F = 0°C$

(b) $\dfrac{5}{9} \times (72°F - 32)$

$= \dfrac{5}{9} \times 40 = 22.22°C$

33. (a) meter   (d) cubic meter
     (b) square meter   (e) liter
     (c) gram   (f) degrees Celsius

35. (a) 8000 ms $= \underline{}\ \mu$s
     $= 8000 \times 10^{-3} = \underline{}\ 10^{-6}$
     $= 8000.\overset{\frown}{000} \times 10^{-6} = 8{,}000{,}000 \times 10^{-6}$ s
     $= 8000$ ms $= 8{,}000{,}000\ \mu$s

     (b) 0.02 MV $= \underline{}$ kV
     0.02×10^6 V $= \underline{} \times 10^3$ V
     $0.\overset{\frown}{020} \times 10^3$ V $= 20 \times 10^3$ V
     0.02 MV $= 20$ kV

(c) 10 km $= \underline{}$
     10×10^3 m $= \underline{} \times 100$ m
     $10.\overset{\frown}{000} \times 100$ m $= 10{,}000$ m
     10 km $= 10{,}000$ m

(d) 250 mm $= \underline{}$ cm
     250×10^{-3} m $= \underline{} \times 10^{-2}$ m
     $25.\overset{\frown}{0} \times 10^{-3}$ m $= 25 \times 10^{-2}$ m
     250 mm $= 25$ cm

37. (a) The base is 2, the exponent 3.
     (b) $2^8 = 256$ cells
39. No
41. $5 \times 4 = 20$ quarts, $20 \div 4 = 5$ gallons
43. 1 liter $= 1000$ mL, $1000 \div 14 = 71$ hours
45. $3\frac{1}{4} - 1\frac{1}{2} = 1\frac{3}{4}$ hp

Chapter 5
1. a
3. d
5. c
7. c
9. b

(The answers to Communication Skill Questions 11 through 25 can be found in the indicated sections that follow the questions.)

27. (a) $4x = 11$
     $4 \times x = 11$
     $\dfrac{\cancel{4} \times x}{\cancel{4}} = \dfrac{11}{4}$ $(\div 4)$
     $x = \dfrac{11}{4}$
     $x = 2.75$

(b) $6a + 4a = 70$
     $10a = 70$
     $10 \times a = 70$
     $\dfrac{\cancel{10} \times a}{\cancel{10}} = \dfrac{70}{10}$ $(\div 10)$
     $a = \dfrac{70}{10}$
     $a = 7$

(c) $5b - 4b = \dfrac{7.5}{1.25}$
     $1b = \dfrac{7.5}{1.25}$ $(1b = 1 \times b = b)$
     $b = \dfrac{7.5}{1.25}$
     $b = 6$

(d) $\dfrac{2z \times 3z}{4.5} = 2z$ $\begin{bmatrix} 2z \times 3z = 2 \times z \times 3 \times z \\ = (2 \times 3) \times (z \times z) \\ = 6 \times z^2 \\ = 6z^2 \end{bmatrix}$

$\dfrac{6z^2}{4.5} = 2z$

$\dfrac{6z^2}{\cancel{4.5}} \times \cancel{4.5} = 2z \times 4.5$ $(\times 4.5)$

$6z^2 = 2z \times 4.5$

$\dfrac{\cancel{3}\cancel{6} \times \cancel{z} \times z}{\cancel{2} \times \cancel{z}} = \dfrac{(2 \times z) \times 4.5}{2 \times z}$ $(\div 2z)$

$3z = 4.5$

$\dfrac{\cancel{3} \times z}{\cancel{3}} = \dfrac{4.5}{3}$ $(\div 3)$

$z = \dfrac{4.5}{3}$

$z = 1.5$

29. (a) Power $(P) =$ voltage $(V) \times$ current (I)
     1500 watts $= 120$ volts $\times\ ?$

Transpose formula to solve for I.

$$P = V \times I$$
$$\frac{P}{V} = \frac{\cancel{V} \times I}{\cancel{V}} \quad (\div V)$$
$$\frac{P}{V} = I$$
$$I = \frac{P}{V}$$

$$\text{Current } (I) \text{ in amperes} = \frac{\text{power } (P) \text{ in watts}}{\text{voltage } (V) \text{ in volts}}$$
$$I = \frac{1500 \text{ W}}{120 \text{ V}}$$
$$\text{Current } (I) = 12.5 \text{ amperes}$$

(b) Voltage (V) in volts $=$ current (I) in amperes $\times$ resistance (R) in ohms

120 volts = 12.5 amperes × ?

Transpose formula to solve for R.
$$V = I \times R$$
$$\frac{V}{I} = \frac{\cancel{I} \times R}{\cancel{I}}$$
$$\frac{V}{I} = R$$
$$R = \frac{V}{I}$$

$$\text{Resistance } (R) \text{ in ohms} = \frac{\text{voltage } (V) \text{ in volts}}{\text{current } (I) \text{ in amperes}}$$
$$R = \frac{120 \text{ V}}{12.5 \text{ A}}$$
$$\text{Resistance } (R) = 9.6 \text{ ohms}$$

31. $x - \$373 = \$1269, x = \$1642$
33. $\$98 - \$48 - \$16.38 = x, x = \$33.62$
35. columns × rows = number of cells, $7 \times x = 112$, $x = 16$ rows
37. $x\%$ of $18 = 43, x = 42$
39. $\dfrac{25 \text{ compressions}}{10 \text{ breaths}} = \dfrac{5}{2} \text{ ratio}, \dfrac{40}{x} = \dfrac{5}{2}, x = 16$

Chapter 6

1. b
3. c
5. d
7. b
9. c
11. d
13. a
15. d

(The answers to Communication Skill Questions 17 through 25 can be found in the indicated sections that follow the questions.)

27. (a) $C = \sqrt{A^2 + B^2}$
    $= \sqrt{20^2 + 53^2}$
    $= \sqrt{400 + 2809}$
    $= \sqrt{3209}$
    $= 56.65$ mi

(b) $C = \sqrt{2^2 + 3^2}$
    $= 3.6$ km

(c) $C = \sqrt{4^2 + 3^2}$
    $= 5$ inches

(d) $C = \sqrt{12^2 + 12^2}$
    $= 16.97$ mm

29. (a) $\sin 0° = 0$
    (b) $\sin 30° = 0.5$
    (c) $\sin 45° = 0.707$
    (d) $\sin 60° = 0.866$
    (e) $\sin 90° = 1.0$
    (f) $\cos 0° = 1.0$
    (g) $\cos 30° = 0.866$
    (h) $\cos 45° = 0.707$
    (i) $\cos 60° = 0.5$
    (j) $\cos 90° = 0$
    (k) $\tan 0° = 0$
    (l) $\tan 30° = 0.577$
    (m) $\tan 45° = 1.0$
    (n) $\tan 60° = 1.73$
    (o) $\tan 90° = \infty$ (infinity)

31. (a)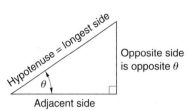

Known values	Unknown values
$H = 27$ miles	$O = ?$
$\theta = 37°$	$A = ?$

We must calculate the length of the opposite and adjacent sides. To achieve this we can use either H and θ to calculate O (SOH), or H and θ to calculate A (CAH).

$$\sin \theta = \frac{O}{H}$$
$$\sin 37° = \frac{O}{27}$$
$$0.6 = \frac{O}{27}$$
$$0.6 \times 27 = \frac{O}{\cancel{27}} \times \cancel{27}$$
$$O = 0.6 \times 27$$
$$\text{Opposite} = 16.2 \text{ mi}$$

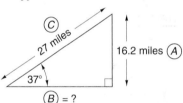

The next step is to use the Pythagorean theorem to calculate the length of the unknown side now that the length of two sides are known.

$$B = \sqrt{C^2 - A^2}$$
$$= \sqrt{27^2 - 16.2^2}$$
$$= \sqrt{729 - 262.44}$$
$$= \sqrt{466.56}$$

Adjacent or $B = 21.6$ mi

(b)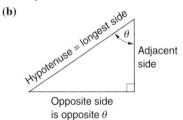

Known values	Unknown values
$O = 29$ cm	$H = ?$
$\theta = 21°$	$A = ?$

We can use either O and θ to calculate H (SOH), or O and θ to calculate A (TOA).

$$\tan \theta = \frac{O}{A}$$

APPENDIX B / ANSWERS TO ODD-NUMBERED PROBLEMS

$$\tan 21° = \frac{29 \text{ cm}}{A}$$
$$0.384 = \frac{29}{A}$$
$$0.384 \times A = \frac{29}{\cancel{A}} \times \cancel{A} \quad (\times A)$$
$$\frac{\cancel{0.384} \times A}{\cancel{0.384}} = \frac{29}{0.384} \quad (\div 0.384)$$
$$A = \frac{29}{0.384}$$
$$A = 75.5 \text{ cm}$$

Now that O and A are known, we can calculate H.

$$C \text{ (or } H) = \sqrt{A^2 + B^2}$$
$$= \sqrt{75.5^2 - 29^2}$$
$$= \sqrt{5700.25 + 841}$$
$$= \sqrt{6541.25}$$
$$= 80.88 \text{ cm}$$

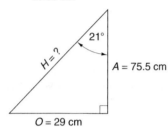

(c)

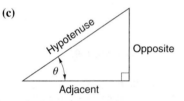

Known values	Unknown values
H = 34 volts	θ = ?
A = 18.5 volts	O = ?

Because both H and A are known, we can use cosine to calculate θ (CAH)

$$\cos \theta = \frac{A}{H}$$
$$\cos \theta = \frac{18.5 \text{ V}}{34 \text{ V}}$$
$$\cos \theta = 0.544$$
$$\cancel{\text{invcos}} \times \cancel{\cos} \theta = \text{invcos } 0.544 \quad (\times \text{invcos})$$
$$\theta = \text{invcos } 0.544$$
$$\theta = 57°$$

To calculate the length of the unknown side, we can use the Pythagorean theorem.

$$B = \sqrt{C^2 - A^2}$$
$$= \sqrt{34^2 - 18.5^2}$$
$$= \sqrt{813.75}$$
$$= 28.5 \text{ V}$$

33. $S = \pi d^2 = \pi \times 2^2 = 12.57 \text{ in}^2, V = \frac{1}{6}\pi \times d^3 = 4.189 \text{ in}^3$

35. $L = \frac{1}{2}ps$
$= \frac{1}{2} \times 24 \times 8 = 96 \text{ in}^2$

37. $b = 27°, a = 63°, a + b = 90°$
39. The plumb bob string should pass through the mid-point mark of the crossbar in the A-frame if the stones are level.
41. $c^2 = a^2 + b^2, 27^2 = 21^2 + x^2, x = 17$
43. $V = \pi r^2 h = 21.21$ cubic feet
45. $25.6 \div 3.2 = 8:1$

Index

Calculator Keys
Add key, 42
Change-sign key, 66
Conversion functions, 40
Convert key, 109
Cosine key, 188
Decimal-point key, 33
Divide key, 43
Equals key, 42
Expand function, 133
Exponent key, 95
Factor function, 133
Inverse cosine, 189
Inverse sine, 187
Inverse tangent, 191
Multiply key, 43
Normal, scientific, engineering modes, 98
Order of operations, 79
Percent key, 56
Pi key, 38
Reciprocal key, 38
Round function, 55
Sine key, 186
Solve function key, 143
Square key, 88
Square root key, 89
Subtract key, 42
Tangent key, 190
Xth root of the Y key, 90
Y to the X power key, 88

A
Acute angle, 172
Addend, 45
Adding fractions, 9
Adding mixed numbers, 14
Adding positive and negative numbers, 67
Addition, 43
Adjacent, 185
Algebra rules, 163
Algebra, 118
Ammeter, 112
Apex, 197
Arabic number system, 32
Arcsine, 187
Associative law of addition, 44
Associative law of multiplication, 49
Augend, 45
Averages, 57

B
Babbage, Charles, 118
Base, 4
Basics of algebra, 120
Borrowing, 18

C
Calculating in decimal, 41
Canceling fractions, 25
Carry, 44
Cartesian coordinate system, 153
Circle, 182
Circle formula example, 144
Circumference, 144
Coefficient, 152
Common denominator, 10
Commutative law of addition, 43
Commutative law of multiplication, 48
Cone, 197
Converting decimal fractions to written fractions, 39
Converting to powers of ten, 95
Converting written fractions to decimal fractions, 35
Cosine of theta, 188
Customary system, 100
Cylinder, 196

D
Decimal calculation, 30
Decimal fraction, 32
Decimal point, 32
Decimus, 32
Degrees Celsius, 106
Degrees Fahrenheit, 106
Delta, 154
Denominator, 7
Descartes, René, 84
Diameter, 145
Difference of perfect cubes, 160
Difference of perfect squares, 160
Difference, 46
Directly proportional, 122
Distributing, 132
Distributive law of multiplication, 133
Dividend, 51
Dividing fractions, 23
Dividing mixed fractions, 24
Dividing positive and negative numbers, 75
Division, 51
Division of exponents, 157
Divisor, 51

E
Electrical conversions, 107
Electrical prefixes, 107
Electrical units, 107
Engineering notation, 97
Equality on both sides of the equal sign, 120
Equation, 120
Equations, 152
Equilateral, 181
Exponents raised to a power, 158
Exponents, 84, 157
Expressing positive and negative numbers, 65
Extra zeroes, 35

F
Factor, 12
Factoring patterns, 159
Factoring, 132
Fermat, Pierre de, 168
Floating point number system, 98
Formula, 120
Fraction bar, 7
Fractional P\power, 158
Fractions, 6
Frustum, 197

G
Gauss, Carl, 2
Geometric terms, 170
Geometry, 168
Graphing linear equations, 154
Graphing, 152

H
Hypotenuse, 185

I
Improper fractions, 13
Increment, 5
Integer, 152
Inverse sine, 187
Inversely proportional, 123
Isosceles triangle, 181

L
Lateral surface area, 196
Linear equations, 152
Literal number, 120, 152
Lowest common denominator, 11

M
Measuring devices, 109
Measuring power, 115
Metric prefixes, 102
Metric system, 101
Metric unit of area, 105
Metric unit of length, 103
Metric unit of temperature, 106
Metric unit of volume, 105
Metric unit of weight, 105
Metric units, 101
Micrometer, 109
Minuend, 46
Mixed number, 13
Multimeter, 112
Multiplicand, 49
Multiplication, 48
Multiplication of exponents, 157
Multiplier, 49
Multiplying fractions, 20
Multiplying mixed fractions, 21
Multiplying positive and negative numbers, 73

N
Negative exponents, 157
Newton, Isaac, 30
Numerator, 7

O
Obtuse angle, 172
Ohmmeter, 113
Ohm's law example, 137
Opposite, 185
Order of operations, 78

P
Parallelogram, 173
Pascal, Blaise, 62
Percentages, 56
Pi, 38

Plane figures, 173
Point-slope form, 156
Polygon, 173
Polygons, 183
Positive and negative numbers, 62
Power formula example, 142
Powers and roots in combination, 90
Powers of ten, 94
Powers of ten in combination, 96
Prime number, 12
Prism, 195
Product, 49
Proportion, 122
Protractor, 170
Pyramid, 198
Pythagorean theorem, 176

Q
Quadratic equations, 160
Quadratic formula, 161
Quadrilaterals, 173
Quotient, 51

R
Radius, 144
Radix, 4
Raising a base number to a higher power, 86
Ratio, 122
Ratios, 52
Reciprocals, 37
Rectangle, 174
Reducing fractions, 15
Remainder, 51
Representing fractions, 6
Reset and carry, 5
Resultant vector, 178
Rhombus, 174
Right triangle, 176
Right-angle triangle, 173
Rise, 154
Root of a number, 88
Root power, 158
Rounding off, 54
Run, 154

S
Scalene triangle, 180

Scientific notation, 97
Second-degree equations, 160
Sides, 170
Significant places, 55
Sine of theta, 186
Sine wave, 191
Slope intercept form, 155
Slope of align, 154
Solid figure, 195
Sphere, 197
Square and square root, 135
Square of a number, 88
Square, 173
Standard form, 152
Statistics, 58
Story problems, 129
Subscript, 4
Substitution, 144
Subtracting fractions, 17
Subtracting mixed fractions, 18
Subtracting positive and negative numbers, 70
Subtraction, 46
Subtrahend, 46
Sum of perfect cubes, 160
Sum, 44

T
Tangent of theta, 189
Theta, 185
Total surface area, 196
Transposing formulas, 136
Transposition, 125
Trapezoid, 175
Treating both sides equally, 121
Trigonometry, 185

V
Vector addition, 178
Vector diagrams, 178
Vernier caliper, 110
Vertex, 170
Vinculum, 7
Voltmeter, 113
Volume, 196

W
Weights and measure, 100
Whole numbers, 4